NAVIGATION FOR YACHTSMEN

Navigation for Yachtsmen

THIRD EDITION, REVISED

Mary Blewitt, F.R.I.N.

STANFORD MARITIME LONDON

Stanford Maritime Limited
Member Company of the George Philip Group
12 Long Acre London WC2E 9LP

First published 1964 for Yachting World
Third Edition 1973
Revised 1974, 1977
© Mary Blewitt 1973

Printed in Great Britain by
The Camelot Press Ltd, Southampton

ISBN 0 540 00963 6

FOREWORD TO THE FIRST EDITION
by Captain John Illingworth R.N. (Ret.)

Navigation For Yachtsmen will prove a useful book, partly because it covers a lot of ground not adequately treated elsewhere and partly because it brings together, within its small covers, a lot of information very necessary to the yachtsman which would take a great deal of digging out of larger manuals on Navigation. By the same token, it discards a great deal of matter which may be necessary to the professional ship navigator but is redundant for the yachtsman's purpose.

It is a forthright book in which the author has not hesitated to give advice, as well as information, when and where she judges this to be useful; advice which she is in a good position to offer, stemming as it does from extensive and practical long-distance sailing on both sides of (and across) the Atlantic. Indeed fourteen further sailing seasons have passed since *Myth of Malham* won her second Fastnet with the author navigating for me.

PREFACE

I have written this book with the aim of helping the amateur navigator. There are, however, no explanations of the technical know-how, since there are innumerable publications where such techniques as plotting a bearing, using Consol, recognition of lights, etc., are fully and most adequately described. The modern navigator can count on much assistance from wireless, charts, books and a number of instruments; yet all of these may fail him if he lacks forethought, care or judgment. There is no ABC which tells him to do such-and-such at such-and-such a moment and all will be well. Each occasion is individual and different from all others (although of course very similar) and only the man on the spot, at the time, can decide what the probabilities are and what should be done. I have tried to make the navigator realise that there are a number of essential questions which he should continually ask himself and be able to answer, such as: Have I tried all the aids available? Has any unnoticed error crept in? Where am I? What shall I do now? Remember, however, that reading this book will not automatically make anyone a good navigator; only experience and practice can do that.

This book is designed primarily for the navigation of sailing yachts, but will serve equally well for motor yachts if the parts solely applicable to sail are omitted. It has been written to help those who cruise as well as those who race. The racing navigator probably knows it all already, but I believe that whoever reads and understands it, with a little practice and experience can offer himself as navigator for a race without fear of disgracing himself even though he realises full well, as most of us realise, that he still has a lot to learn.

I am most grateful to Admiral Mario Bini for his criticisms and for his article on compass deviation which I have included as an appendix. It is a subject which is most forbidding to the amateur, and it has been a great pleasure to me to find, at last, that it is possible for those not technically minded to understand something about it.

Preface

I do not flatter myself that anything in this book is original. My thanks are due to John Illingworth, Myles Wyatt, Michael Richey, Dick Scholfield and to many others with whom I have had the fortune to sail, and by whose knowledge and experience I have benefited; and especially to my late husband for his help and advice. I have merely tried to collect and co-ordinate what I have learnt over the years and put it at the disposal of others who wish to master the 'haven-finding art'.

MARY BLEWITT

CONTENTS

1 EQUIPMENT   1

2 PLANNING AND PREPARATION   22

3 AT SEA   30

4 STRATEGY   56

5 REACHING YOUR DESTINATION   69

6 GOING FOREIGN   75

APPENDIX   80

INDEX   99

CHAPTER ONE

EQUIPMENT

COMPASSES

The magnetic compass is by far the most important piece of navigational equipment aboard a yacht, and it is not safe to go to sea without one. If you are without a log, a guess can be made at the speed; if you have no chart, a map *can* be used, or at least a guess made at the direction of land; but with no compass a boat can be well and truly lost. I do not go so far as to say that it is necessary to carry one rowing ashore in Cowes Roads, yet it is easy to imagine a night with fog when a compass might be extremely useful even there.

THE STANDARD COMPASS

This is the name given to the principal compass aboard, which is mounted as far from any magnetic influence as possible and to which all courses are referred. In a small yacht, where there is no room for two compasses, the standard compass is also the steering compass. Aboard large yachts the standard compass is often placed forward of the deckhouse, amidships under the boom, and is usually flat-topped, with an azimuth-ring for taking bearings and a shadow-pin for reading the azimuth of the Sun. A good strong cover is necessary against blows from sheets and other damage. This compass is most useful to the navigator but by no means essential. It is interesting to note that someone as experienced as Rod Stephens, in his own 32ft water-line sloop, *Mustang*, gave such importance to the accuracy of the compass that he had a standard compass (identical with the steering compass so that parts were interchangeable in case of damage) mounted on a binnacle aft of the cockpit.

THE STEERING COMPASS

Separate standard compasses being rare aboard yachts, however, we need only consider the smaller boat where one serves both as standard and steering compass.

Buying a Compass   If you have to buy a compass for a new boat, or put a new compass into an old boat, buy the best. The most suitable modern compass for a small yacht has a magnifying glass dome, such as the Plath, Sestrel or Danforth, and is marked in 360°; it has a minimum of parallax because the lubber line is cardanically suspended and therefore always stays vertical; it is easy to read and should be fitted with a central shadow-pin.

Mounting a Compass   Various considerations come into play when choosing the position for the compass. In a newly designed boat the designer will have provided a suitable place for it; but if you have to move it because of changing from wheel to tiller steering, or vice versa, or for some other reason, there are various points to be kept in mind. As will be seen in the appendix, certain deviations of the compass, if they exist, can only be eliminated by moving the compass or by moving the iron which causes them. In addition, a position too close to the engine may bring difficulties with heeling error. It must be remembered, however, that the disturbing effect of iron diminishes with the cube of the distance, so that a few inches difference in the position of the compass may make the difference between a big error and none at all. The suitability of the chosen position from the 'magnetic' point of view, therefore, must be checked *before* the compass is irrevocably installed.

The compass should be mounted amidships in the best position for the helmsman. Do not say 'this is the best place for the compass but it will get in the way of the main sheet so we'll have to put it elsewhere'; it is the main sheet that must be moved. Make sure that the lubber line is exactly on the fore-and-aft line of the boat; check that there is no difference in heading when the engine is running (a compass too close to the dynamo can be affected by it); and, finally, make sure that it is placed near enough to the helmsman. I have seen boats where the compass, though perfectly visible on a clear day, or to someone with perfect eyesight, cannot be seen through spectacles streaming with rain or spray.

Swinging, Adjusting and Checking a Compass   Swinging the compass and adjusting it are two different and quite separate operations. The compass is swung, first, before adjustment, to find out what deviations there are, and, secondly, after adjustment, to draw up a compass-deviation card if necessary. The compass is adjusted, by installing magnets, to eliminate or reduce certain deviations. Only a

2

professional can have the experience needed to write clearly, accurately and simply enough to explain such a seemingly difficult subject and I am fortunate in being able to include an appendix on the subject by an expert. Admiral Bini, of the Italian Navy, is a Fellow of the Royal Institute of Navigation, a hydrographer and a yachtsman; he has swung and adjusted compasses aboard aircraft-carriers and yachts, and his explanations are both simple and lucid. The appendix explains what various types of iron do to the compass, how their influence is identified and isolated, how various deviations are eliminated, together with practical tips as to how to set about it. Although a careful study of the appendix will enable the navigator to swing and adjust his own compass (except heeling error), many people may still prefer to call in the professional adjuster, either from lack of time or from lack of self-confidence. Even so it is worth their while to read the appendix so that they have some idea of what the adjuster is doing, and have some control as to whether he has done it well or not.

Since the navigator must have absolute faith in his compass, he should check it fairly frequently (in a steel ship the magnetic compass is often checked twice a day). This is particularly necessary if anything unusual happens to the boat: e.g. struck by lightning, shipping abroad, collision, etc. In one boat shipped to the Argentine for the Buenos Aires–Rio race, the navigator checked the compass and noticed that it stuck at a certain point. When taken to pieces it was found that it had been broken by jolting during the voyage, and that therefore the needles could not swing freely. This might easily have gone unnoticed and been the cause of serious errors.

A friend of mine once navigated one of the best known American yachts in a transatlantic race. Assured that the compass was without error, and knowing that the boat had recently been racing, he accepted this and did not check it. After two days in fog, during which he found himself consistently and seriously off course for no ascertainable reason, he was able to get an azimuth of the Sun and found a 15° error. It then emerged that the magnets were altered by means of screws, and that one of the crew, before the race, had gone round with a screwdriver tightening every screw on board including these. The moral is: never trust what others tell you about the compass, make sure for yourself. If you have one of these compasses, incidentally, cover the adjustment screws with tape to ensure their not being altered.

A rough check can be made with the hand bearing-compass. Take this aft, amidships (be careful not to get too close to the backstay) and line it up with the fore-and-aft line of the ship. This can either be done by lining up the mast with the centre of the pulpit, or, best of all, by sighting along the two lubber lines of the steering compass. Then read the hand bearing-compass, calling it continuously for several seconds, while another watches and checks the steering compass. This should be done on various headings if possible, as for swinging.

A more accurate check, if you have a shadow-pin, is made by measuring the azimuth of the Sun when it is low in the sky. Take the time (accurate to the nearest 10 seconds or so) and read where the shadow of the pin falls on the compass card. This 'sun-dial' reading, plus or minus 180°, is the compass azimuth of the Sun at that moment. This figure is then compared with the true azimuth of the Sun for that time and place, corrected for magnetic variation. The working is shown in Table 1.1. If you have no tables there is one in *Reed's Almanac*; it is complicated and needs a clear head and blind faith, but gives the answer.

This method of checking is particularly useful at the end of a long passage when you feel that it would be comforting to know, before making landfall, that the compass has no unknown error.

Table 1.1

B.S.T. 09 h 33 min 20 s	'sun-dial' reading	294°
G.M.T. 08 h 33 min 20 s		−180°
	compass azimuth	114°
From *Almanac* G.H.A. Sun for 0800 12 July 1977		298° 37'
G.H.A. increment for 33 min 20 s		8° 20'
G.H.A.		306° 57'
Assumed longitude		−2° 57'W★
L.H.A.		304° 00'

★East add, West subtract to make whole number nearest your position.
Declination 22° N (nearest whole number of degrees).
Assumed Latitude 50° N (nearest whole number to your position).

rom *Tables* (A.P.3270): Vol. III — Find page for Lat. 50°, Dec. 22° (SAME NAME as both North), read off against L.H.A. 304° under Dec. 22°, Azimuth in column Z, 102° (correct Z to Zn if necessary as explained at bottom of page of tables) correct for magnetic variation. Variation 10° W. 102° + 10° = 112° = magnetic azimuth.

Compare this figure of 112° with the compass azimuth, and the difference is compass deviation. In this case the deviation, on the heading being steered at the time, was 2° W.

Steel Yachts   Any form of iron or steel is of its very nature likely to deviate the compass; when the hull is constructed of or incorporates such materials—steel or ferrocement hulls, steel frames—particular care must be used in siting and compensating the compass, and you must remember that below decks any compass will be unreliable unless it is only a repeater. You should be even more meticulous over checking your compass when sailing in a steel hull than in one of wood, fibreglass or aluminium.

Aboard large steel yachts you will need a compass with a binnacle which enables a complete compensation to be carried out. The compass must be frequently checked, especially if the boat is left for any length of time on the same heading (in harbour, or even during a long passage), or if considerable changes of latitude have been made. It is strongly advisable in a steel yacht of any size to have a gyro-compass and to use the magnetic compass as a reserve. New equipment which can provide a solution consists of a magnetic sensor feeding to repeaters placed where required.

Taking Care of a Compass   Your compass is mounted, swung, adjusted and checked and now you must treat it with respect. Make certain, for example, that if, with an old compass, there are magnets screwed on to the surrounding wood these are not moved, or put back to front, when varnishing. Magnets must not be allowed to rust as this alters them; they should be dipped in grease or oiled before insertion in the compass. Cover the compass when not in use, especially in hot sun. See to it that it is well lit at night and that some form of lighting is available if the ship's batteries fail; it is not pleasant to have to peer at the compass by the light of a failing torch held fitfully by another member of the crew. Above all else be careful what you put near it.

Before the war, a large yacht, running down from the Fastnet to the Bishop Rock, was lucky not to be lost when she found high land ahead of her, in bad visibility. She was well up the Bristol Channel, because someone had stowed a pair of wire sheets under the compass. Today, radio bearings should alert the navigator to the fact that something is desperately wrong, but the situation is to be avoided. Nothing made of iron must be put near the compass, and this applies not only on deck but below deck (under the companion-way on top of the engine) if the compass is mounted at the forward end of the cockpit. Nor is iron the only danger; light-meters for cameras, or cameras with light-meters incorporated, must be kept well away, and transistor wireless sets must on no account be put nearby, since they can cause error from as far as three feet away.

In case the reader thinks that it is unnecessary to pay so much attention to the compass and that it points more or less in the right direction anyway, let me quote from the Race Analysis of the 1963 Hook Race, which should serve as a warning. Two skippers' remarks are: 'Compass fell down. Deviation table probably incorrect'. (I would say that it is extremely unlikely that the compass fell down on its job, someone who should have checked it fell down on his job; or, possibly, one of the helmsman's errors discussed in Chapter 3 came into play.)

'Compass wrong due to stowage of life raft on deck.' The steel compressed air bottle evidently was the culprit.

Before finishing this section on the steering compass let me stress, once again, that the compass *must* be accurate, the navigator *must* be able to have complete faith in it, otherwise all his efforts are in vain.

THE HAND BEARING-COMPASS

In addition to the steering compass you will need a hand bearing-compass for taking bearings. There are various good makes on the market. Remember, if batteries are required for illumination at night, to carry spares, and do not wait for a dark, wet night to find that the old batteries have gone bad and that the whole compass needs cleaning, and the connections sandpapering, before you can get a light out of it.

Miniature hand bearing-compasses are now popular. The beta light permits the elimination of batteries and the newest, neatest compass is a disc four inches across and an inch deep which you can carry in your pocket or round your neck and use to check your bearings when at the helm without wondering where you can lodge the hand bearer safely when not in use.

You cannot expect to take excellent bearings with a hand bearing-compass immediately. It is one of the easiest instruments to use, but do practise before you rely on your bearings. If you find it difficult to keep steady with one hand, hold the bowl in both. Use it in harbour when you know exactly what the bearings are from the chart. Try practising also at night. The ship's movement makes the compass card swing, and while by day you can take the mean, or read it when steady, at night, if you do not keep the light lit, it is easy to read it at the extremity of the swing and therefore have a big error. Occulting lights are easier than flashing lights, but a group flash of three or four gives you a chance to get a good bearing. There are, however, lights which are virtually useless for yachts: the Lowestoft lighthouse flashes once every 15 seconds, and I defy anyone in a small boat, with a bit of a sea running, to get a good bearing on it. Fortunately such lights are few and far between.

Tell-Tale Compass   This is the name given to a compass placed below deck so that the navigator can see what is happening without coming on deck. These compasses are often made with the card reversed so that they can be installed upside down over the navigator's or skipper's bunk. Easy to install, and reliable in an all-wood yacht, they should be sited with the greatest care in a composite boat with steel frames, and are probably unusable in a steel yacht.

Magnetic Variation   Unfortunately, even if the compass has no deviation at all, it still does not point towards the North Pole, but towards the Magnetic North. The difference between true and magnetic north, measured in degrees, is called Magnetic Variation. In some languages, and therefore on some charts, it is called Magnetic Declination. On Admiralty charts a magnetic rose is shown inside the true rose, so that you can read off true, or magnetic, bearings as you wish (provided the magnetic rose is not out of date); most foreign charts do not carry this rose. On all charts, however, the variation is written inside the true rose, with the date. Variation differs from place to place and changes slowly in the course of time. For instance the rose south of the Nab Tower, from a chart of 1974, reads: 'Varn. 7° 00' W (1974) decreasing about 4' annually'. From this you can work out that the variation in 1977 is 6° 48', so that you will still use a variation of 7°. Remember that, unless otherwise specified, bearings are given as true bearings: if you see on the chart 'lights in line 42° 00'', this means that the bearing is true; if the bearing were magnetic it would be written: 42° 00' (M).

If you use up-to-date Admiralty charts and parallel rulers or such (see p. 12), there will be few occasions when you need to convert true to magnetic bearings or vice versa; if, however, you are using navigational set-squares, oldish charts or foreign charts this will always be necessary. It is not as difficult or troublesome as it seems at first sight. Since you are sailing in one area the variation will not change from easterly to westerly and back again, so that it becomes automatic within a very short time to apply the local variation correctly. In the appendix the following rhyme is suggested as a mnemonic:

> Variation west compass reads best,
> Variation east compass reads least.

That is to say, with an easterly variation the magnetic bearing will be less than true, with a westerly variation the magnetic bearing is always greater than true.

It is good practice, even when not necessary, to work from the true rose on the chart. The habit is thus formed of converting rapidly and accurately, so that when, say, in the middle of a stormy night you are confronted with true bearings in the pilot book, you will have no hesitation about transforming them quickly into magnetic bearings. No amount of examples that could be given can take the place of practice on the chart; and to repeat: the moment when you *have* to convert will not come on a lovely sunny day with all the time in the world, but in some emergency, when you are already tired and, if you have had no practice, it seems incredibly difficult and complicated.

Compass Deviation   If the compass has a residual deviation after adjustment (which should be avoidable on a wooden yacht), a compass deviation or correction card must be drawn up, giving the corrections to be applied on the various headings. This really is a bore since it means that every course (not bearings taken with the hand bearing-compass) has to be corrected, both that given to you by the helmsman and vice versa, and the correction will be easterly on some headings and westerly on others. However, if you do happen to be sailing on a boat where there is a large compass deviation error (less than one degree can always be ignored) again remember the rhyme (slightly changed):

> Deviation west compass reads best,
> Deviation east compass reads least.

If you have to correct both for deviation and variation, similar names (both east or both west) are added, if the names are different (one east and one west) the smaller is subtracted from the larger and the name of the larger given to the remainder; i.e. with variation 10° W and deviation 2° E the resulting total correction is 8° W.

CHARTS

Next in importance to the compass is the chart. Charts are made, printed and published by the Admiralty Hydrographic Department and sold through Admiralty Chart Agents. The names and numbers of the charts are given in the *Admiralty Chart Catalogue*, which is available for consultation at any chart agent when you buy your charts. There are two editions, one for the whole world and one for home waters. Study the appropriate catalogue carefully, and give yourself plenty of time to make sure you have bought all the charts necessary.

METRIC CHARTS

A year or two ago measurements on British charts were in feet and fathoms, now a gradual change to metres is being made. As far as U.K.

waters are concerned the first metric charts were of the West Country and the Hydrographic Office is gradually working east. Therefore the very first thing to do with any chart is to *look* to see if it is in metres or fathoms. It need not worry you once you have found out. One fathom or two metres is minimum water for most small yachts—certainly for sailing if not for harbours—with one-and-a-half fathoms or three metres you have enough for any but very deep-keeled or very large boats. During the transition period it might be as well to note down somewhere handy the minimum depth into which you want to go in fathoms and feet, feet, and metres. This should avoid confusion. Conversion tables are given at the side of the chart.

Since 1973 tidal information has been given in metres or in metres and feet and this will make things difficult if your chart is still in fathoms; however it is a transitory period which will only last a few years.

Now planning a cruise from Cowes to the Channel Islands you will want:

> 2675 *English Channel, Eastern Portion* (scale 1 : 500,000)
> This is a small-scale chart for planning your trip, laying off cross-Channel courses and generally getting a bird's-eye view.
>
> 2045 *Christchurch to Owers* (1 : 75,000)
> A medium-scale chart for getting you in and out of the Solent.
>
> 2669 *Channel Islands and Adjacent Coast of France* (1 : 150,000)
> This medium-scale chart will serve for your daily sailing round the islands.
>
> Large-scale charts of all islands and harbours you wish to visit, or where you might possibly have to take refuge: Cherbourg (a westerly gale may force you in there), Alderney, St. Peter's Port, St. Helier, etc. and, of course, your home port.

Do not be stingy about charts. It is no good saving £2 and then not being able to take shelter from a gale in a convenient nearby harbour because you were too mean to buy the chart. If you are sailing along a coast you should carry all the harbour charts of that coast, just in case.

Keep your charts up to date. Old charts can be misleading; not only lights and buoys change, but even channels in such areas as the Thames Estuary, where the banks are slowly but continuously shifting. Charts have the date of publication under the title and, in the left-hand corner, at the very bottom, the date of the latest corrections. Admiralty chart agents have a correcting organisation for which you pay a small fee for each chart. Corrections are issued in the *Admiralty Notice to Mariners* which is available at any Customs House. It should be consulted before sailing to see if there is anything new in the area to which you are going. Many of these notices are not interesting, but consider the following:

THE SOLENT APPROACH:
Nab Tower. Dredged channel northward; light buoys
established. A new channel . . . . marked by three
light buoys . . . .

This means that for some time there were three light buoys which were not on your chart, in an area of water you were very likely to cross. At the best you were going to waste valuable positioning assistance, at the worst you were going to get appallingly confused if you did not know about these lights. So give a glance at the *Notices*, especially, I would say, for waters much used by big ships, since it is here that variations will probably be found. You may also find warnings of a temporary kind; that such-and-such a light vessel has been withdrawn for repairs, or that a certain buoy has been cut adrift by a collision and has not yet been replaced.

Charts contain a marvellous amount of information, but many people do not know how to get the best out of them. I spent a long time, when I started to navigate, avoiding wrecks marked 'P.A.' which I had decided meant 'partially awash'. Moving out of the North Sea into the deeper waters of the Channel it did eventually occur to me that some of these wrecks were in too deep water to be ever awash, and then I found that it meant 'position approximate'. This kind of stupidity is merely wasting the information supplied to you. Take a medium scale chart, get a list of the abbreviations (*Reed's* or Admiralty Publication No. 5011) and settle down to study your chart. If you think you know everything already, ask yourself the following questions:

What do the signs Ⓐ Ⓑ mean and how do you use them?

What does ⅏ mean?

What does F.Fl. mean?

What is the difference between ⊥ and + ?

What is the difference between 5 kn→ and 5 kn→ ?

What does 6 mean?

Not all the signs and abbreviations on the charts will be useful to you, but you should know all those likely to affect the yachtsman. Hence, study carefully the symbols for different buoys, so that by looking at your chart you know what to expect; wreck buoys, pillar buoys, spar buoys, etc. Get someone to ask you questions from the chart so that you eventually feel confident that you are getting the maximum benefit of all the information on it.

LIGHTS, BUOYS, FOG SIGNALS

The information given on the chart about these navigational aids is not complete, and further details are supplied in the *Admiralty List of Lights, Reed's Almanac* and *Brown's Almanac.* The same publications give descriptions of the different types of fog signals and lights (flashing, quick flashing, interrupted quick flashing, etc.) which may be heard or seen, and of wreck marks. All this should be read and studied by the would-be navigator *before* going to sea; he should not have to search for the meaning of an abbreviation in the middle of the night in a Force 7 wind.

The entire buoyage system in U.K. and North European waters will be changed between 1977 and 1980. This means that your charts and books will be out of date from time to time. The dates of changes in various areas are listed in *Reed's* and elsewhere, but the unfortunate navigator must for some time expect what he finds and not to be too surprised if a black starboard hand buoy has turned green or a green buoy has stayed black. The new system is logical and should, when fully installed and understood, be better than the old, but extra care must be taken while both systems are still in use.

ROAD MAPS

New constructions on land take time to get on to a chart. Since they are not corrections which can be added yearly, they require either a new edition of the chart or even a new survey. It is for this reason that it can be very useful to carry up-to-date road maps of the coast you are visiting, scale 1 : 200,000 or thereabouts. Many villages, marked on the charts as 'houses and church', have turned into long strung-out bathing centres, which at night are a blaze of lights. This can be most disconcerting; you know where you are (you think) when you see what appear to be the lights of a large city. On the chart there is nothing of that size for miles. Have you made a mistake? Can you trust your dead reckoning? A glance at the road map shows you a string of houses along the coast road and so removes that doubt from your mind.

Near Sète, in the south of France, there is a refinery, that, recently, was not on the charts. Making landfall at Sète this refinery is the first thing you see, especially at night, since the flares are higher and brighter than the Sète light. The refinery, however, was marked on the Michelin map, and so the navigator's job was made much easier. Maps must be

used with great caution and must *never* be used for laying off bearings
or measuring distances.

CHART LOCKERS

Charts must be kept flat and not rolled. There is no more infuriating
thing aboard ship, to my mind, than trying to find, extricate and then
use a chart from a large roll. It is better to fold them again if you have
no space large enough to take them. While the majority of charts
can live under a mattress and be got out in harbour, or when it is calm,
the few that are immediately necessary must be found a home. If you
have a clear bulkhead, a very neat locker can be made, into which the
charts can be put standing up. The names of the charts can be written on
the top edge and any chart can be extracted easily.

CHART INSTRUMENTS

Your three main jobs on the chart are laying off bearings and courses,
measuring distances and plotting your dead reckoning. For this you
need parallel rulers, or equivalent, and dividers.

PARALLEL RULERS

There are various schools of thought about the best method of transferring a line across a chart so that the new line is parallel with the
original one. This has to be done by some form of ruler and traditionally
has always been effected by parallel rulers. Some have rollers underneath,
some are cut in half and the halves joined by a movable lattice. Parallel
rulers may be all right on a large, perfectly flat chart table in the
Queen Elizabeth II, but to my mind they are unsuited for use on a small
boat. If the chart is slightly damp in one place, or has been folded so
that there is a crease, or if there is the slightest irregularity under the
chart, they slip out of parallel with the greatest of ease. And when you
have moved them across the whole chart how can you be absolutely
sure that they have not slipped? There are, however, two other ways in
which a line can be ruled on the chart at a certain angle without 'moving the line' across it.

PLOTTING INSTRUMENTS

Fortunately this dislike of parallel rulers is now common and a number
of simpler and better methods are available. Among these there is for
instance the Hurst plotter, produced by Brookes and Gatehouse
(Fig. 1.1). A disk marked in 360° is mounted on top of a gridded
perspex square. The disk can be clamped at any required position to
read the magnetic variation. The grid is lined up with the nearest

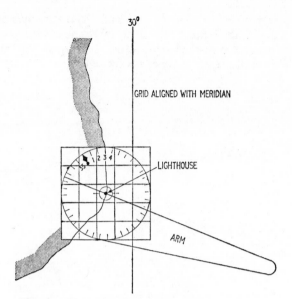

Fig. 1.1   Hurst plotter showing bearing of 325° (M) with
30° of westerly variation

meridian or parallel of latitude on the chart and the arm is used to read
off courses or to draw bearings. To lay off a bearing from a lighthouse,
for example, centre the rose over the lighthouse, line up the grid and
then swing the arm and draw in the bearing. The magnetic variation is
automatically taken care of. These plotters have the advantage that no
ruler need be moved over the chart, but have the considerable dis-
advantage that the area of the chart covered by the perspex square is
unusable and this when one is near to the lighthouse brings its own
plotting problems. I personally use this plotter but keep a pair of
navigational set-squares handy for close or very accurate work.

NAVIGATIONAL SET-SQUARES

These are equilateral, right-angled set-squares with a rose drawn from
the centre of the hypotenuse. They are used by the German and Italian
navies, and elsewhere in Europe; and are now on sale also in the
United Kingdom. If you can get them, and get into the habit of using
them, they are probably as good as or better than anything else. There
is only one snag, you must use true bearings. Fig. 1.2 shows how to
lay off a bearing from a lighthouse. The bearing of the lighthouse was

126°(M), magnetic variation 8°W, 118° true. Lay the set-square so that the meridian cuts the centre of the hypotenuse and coincides with the line on the set-square marked 118°. Move the set-square up or down the meridian until the hypotenuse cuts the lighthouse and draw your line.

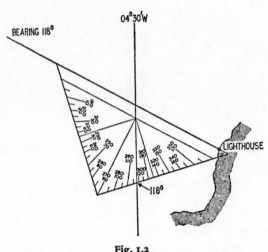

Fig. 1.2

In Fig. 1.3 the lighthouse bears 008° (M), 360° true. The hypotenuse (360°, 180°) is aligned along the meridian and the set-square is slid along the other set-square (which merely does the job of a ruler) until it cuts the lighthouse.

In Fig. 1.4 the course to be measured is *AB* (or *BA*). The hypotenuse is laid along the course and moved until the centre coincides with the meridian. The course is then read off from the meridian, *AB* 254°, *BA* 074°, magnetic variation 8°W, compass deviation 2°W, total correction 10°W; course 264° (M) or 084° (M). As can be seen, one set-square is usually enough, but they are sold in pairs. If you are prepared to use parallels of latitude as well as the meridians it is hardly ever necessary to use two, but 90° must be added or subtracted.

You may object that it is too troublesome to correct for magnetic variation but remember that working in one area, with a constant variation, it becomes automatic and no other method allows you to lay off bearing and courses with such speed and accuracy.

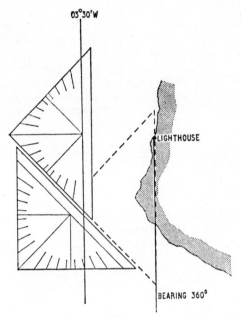

03°30'W

LIGHTHOUSE

BEARING 360°

Fig. 1.3

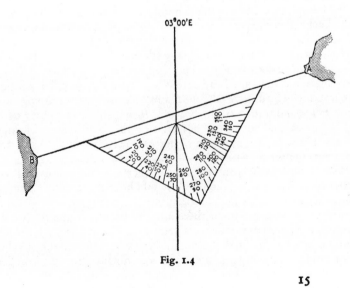

03°00'E

A

B

Fig. 1.4

DIVIDERS

Buy a pair of strong, well-balanced dividers. There are those which cross over, and you squeeze the ring at the top to separate the points; and there are those which are hinged at the top. I use a pair of the latter, light plastic with non-rusting points, but only because I was given them; I do not think there is any intrinsic advantage in either kind. The important thing is to get used to using them quickly and accurately.

ELECTRICAL EQUIPMENT

WIRELESS

There are a number of good sets on the market for use aboard yachts, for weather forecasts, radio bearings and consol. I have not got the technical knowledge to recommend one rather than another, but any reputable ship's chandler will tell you the pros and cons of the different kinds. Some, such as the Heron, have a small compass attached to the aerial so that the bearing can be read directly; others have no compass and the bearing is read so many degrees to port or starboard of the ship's head, and the ship's head checked from the steering compass. If you have one of the former type you should check the small compass on various headings, holding it as you would to take a bearing; take care not to hold the compass too close to your face, and therefore to the headphones, as the latter can induce an error in the compass. In small yachts the wireless should be independent of the ship's batteries.

Radio Bearings   It is not the object of this book to repeat the excellent technical instructions in *Reed's* and the *Admiralty List of Radio Signals* for using radio beacons. Read carefully what is said in either of these publications which give the positions, frequencies, grouping, call signs, range and timing of the beacons. *Reed's* goes into greater detail for beginners and gives the beacons for the North Sea, Channel, Atlantic coast of Europe and the western Mediterranean; the Admiralty list covers the whole world.

With a steel yacht you can have neither a compass nor an aerial below deck. And the aerial on deck should be in a permanent position where it can be calibrated. Instructions for doing this are given in *Reed's*. On a wooden yacht there are few problems, the set (or the aerial if it is separate) must be kept away from metal, especially standing rigging. *Reed's*, discussing the type where the aerial is attached to a hand bearing-compass, says: 'The safe distance from an iron keel or engine weighing one ton is approximately 5ft; 4 tons, 10ft'. If you have an iron keel, therefore, it is probably better to buy the Beme type, and read the bearing from the steering compass; otherwise you must always take the bearing on deck.

Consol   This is a long-range radio navigational aid for which special charts are provided. The charts are listed in the *Admiralty Chart Catalogue*. Instructions for using them are found on the charts themselves, in the *Admiralty List of Radio Stations* (Vol. V) and in *Reed's*. The Consol stations can be picked up on an ordinary ship's wireless as they are broadcast on the same frequency band as the radio beacons. Consol covers the North Sea and east Atlantic as far as Gibraltar.

Weather Forecasts   *Reed's* gives the frequencies and times of U.K. and Continental shipping forecasts; the *Admiralty List of Radio Signals* (Vol. III) gives detailed particulars of meteorological services throughout the world.

ECHO SOUNDER

Echo sounders are extremely useful aboard and are an important safety factor in fog or bad visibility. I consider that all yachts should carry one. After it has been installed the echo sounder must be adjusted; you can set it either to read the depth of water under the keel so that the boat will ground at zero, or to read the depth of water from the surface so that, if the boat draws 6ft, you will ground at a reading of 6ft. The latter is preferable as the depth is more easily compared with the charts. In our own boat I put a red line on the dial at 7ft so that everyone could see clearly when we were getting too close in. The echo sounder is adjusted in harbour by comparing its reading with that of an accurately marked lead and line.

OTHER EQUIPMENT

LOG

As far as I know there are no small ship's logs in *general* use in Europe other than Walker's. Read the instructions for the length of line, and carry a spare spinner in reserve. *Artica II* lost two spinners on the Brest-Teneriffe race, which were said to have been eaten by sharks! Logs are fairly tough instruments and need only a little oil occasionally for maintenance.

Excellent but expensive log and speedometer combinations are available for racing boats, where the drag of a conventional log is considered too great, but too much reliance should never be put on small electronic instruments and a Walker-type log should be carried aboard always.

LEAD

All yachts must carry a lead. 10 fathoms of line should be long enough for any but the largest boats.

BAROMETER

You should have a barometer. A barograph is better still, but it is expensive and takes up a lot of room. It also needs a damper to stop the movement of the boat blurring the reading. The barometer often gives the only warning of bad weather you will get.

STOP-WATCH

There should be a stop-watch on board for timing lights at night.

SEXTANT AND DECK WATCH

These are only necessary if you wish to use celestial navigation.

BOOKS

ALMANAC

Two almanacs are published in Great Britain every year, *Brown's* and *Reed's*. Both contain the ephemerides, tidal information, lights, beacons and buoys of the United Kingdom and N. Europe and a great deal of other information useful to the navigator. One or the other is essential, but they do not quite fulfil the same needs.

Reed's is mainly for use in North European waters and has useful explanatory chapters and other general information. It combines the *Admiralty List of Lights*, *Admiralty List of Radio Signals* and *Admiralty Tide Tables* for this area. It obviates the need for the last two, but its light list, unfortunately, is less easy to consult than the **Admiralty list**.

Brown's is the professional's almanac. It gives world coverage on tidal information, lists of distances between ports for all the world routes, gives the Atlantic steam routes and a great deal of legal information pertaining to cargoes, salvage, victualling, etc. It presumes that you have the *Admiralty List of Radio Signals* aboard and does not give any radio beacons. It is less adapted for sailing round the British Isles, but would be more suitable than *Reed's* for a 'round the world' trip.

TIDAL STREAM ATLASES

Admiralty pocket atlases are published in eleven volumes: *English and Bristol Channels*; *Solent and Adjacent Waters*; *Approaches to Portland*; *The Channel Islands and the Adjacent Coast of France* and others round the coasts wherever tides run strongly. Stanford Maritime publish English Channel atlases in three volumes. They include similar information to those of the Admiralty in an easier form and based on additional hydrographic data, and also give useful tables for predicting tidal heights at various ports. These will replace portions of *Reed's* where the tidal stream maps are small. The *Stanford's Tidal Atlases* are new at the time of writing but will, I believe, prove very popular.

In all tidal atlases streams are given hourly at high water Dover and for the 6 hours before and after, and for selected portions of the U.K., Channel and North Sea coasts. Arrows show the tidal streams and are marked with the speed of the streams for mean neaps and mean springs.

ADMIRALTY LIST OF LIGHTS

This is published in twelve volumes covering the whole world. If you are outside the range of *Reed's* or *Brown's* you must carry the appropriate volumes. Full descriptions of the lights, abbreviations used, fog signals, etc., are included together with illustrations of the various light characteristics and a discussion of the factors affecting their use. Many navigators will prefer this publication to *Reed's* as the information is laid out in very accessible form and is consistent with that used on charts.

ADMIRALTY LIST OF RADIO SIGNALS

Vol. I Communications.
Vol. II Navigational Aids: D.F. Stations and Radio Beacons.
Vol. III Radio Weather Messages; Meteorological Codes.
Vol. IV Meteorological Observation Stations.
Vol. V Radio Time Signals, Standard Frequency Services, Standard Times, Radio Navigational Warnings (with Ice Reports), Position Fixing Systems.

These volumes are usually published bi-annually. You will not need them if you keep within the area covered by *Reed's*. Going further afield you should have Vols. II, III and V. Each volume gives a world-wide coverage of its subject.

Good Radio Beacon diagrams and Radio Information charts published by the Admiralty are also available.

THE PILOT

The Admiralty Hydrographic Department publishes *Pilots* of the whole world in a very large number of volumes. The great work has been compiled over the years, one can almost say centuries, and contains an unending source of information and interest for the navigator. Each volume starts with an article on general navigation (navigational publications, use of charts, and general remarks). This is by no means concerned only with large ships as some might expect, but is an extremely succinct exposition of navigation and the dangers involved, and can be read with advantage by all save professional navigators. There follows an equally clear chapter on General Meteorology. The introduction to the particular volume follows, giving details of winds and currents and tides, frequencies of gales and other information of the coasts discussed. Then comes the main section of the *Pilot* describing the coasts, anchorages and harbours in detail. When sailing to an area you do not know well the appropriate *Pilots* must be carried and studied. 'Local' waters are covered in several volumes, updated periodically by *Supplements* and *Notices to Mariners*. These, and the *Pilots*, are available through Admiralty chart agents.

THE MARINER'S HANDBOOK

Although the last part of this book is about forms of ice that one hopes never to meet, it contains much valuable information on charts and publications, the use of charts and other navigational aids, navigational hazards, tides, currents, and general maritime meteorology. It is an enjoyable book to have aboard.

NAVIGATING IN OTHER PEOPLE'S BOATS

If you are asked to navigate a boat with which you are unfamiliar, what can you reasonably expect to find on board and what should you bring yourself? You are justified in expecting to find on board (for the British Isles and environs) the following ten items, the first *six* of which are essential:

1. *Compass and deviation card of recent date*—Check the compass on as many headings as you get the chance to, and, if you find big errors, I consider you are justified in refusing to sail until it has been re-swung.

2. *Hand bearing-compass*—Check lights and spare batteries.

3. *Log*—Put some oil in it, if you have doubts about maintenance, and check the length of line. Large yachts need a 60ft line and an extra sinker.

4. *Lead and line.*

5. *Reed's Almanac*—(and/or necessary Admiralty volumes) if in doubt take your own copy.

6. *Charts*—All those necessary, corrected up to the beginning of the season.

7. *Echo sounder*—Not essential; if on board check setting.

8. *Wireless*—Not 100 per cent essential but found on all yachts today. Make sure that there are spare batteries.

9. *Barometer*—Not essential; but keep a sharp lookout for the weather if you are without one.

10. *Log book*—A book is not essential; any bit of paper *will* do, but make sure there is at least that.

The following four items you should bring with you:

1. *Tidal Atlases*—You can do without these by using the chart and *Reed's*, but it is not a bad thing to have your own.

2. *Rulers or Set-squares or Sestrel 'Navigator'*—Take your own, which you are used to. On board you may find a very stiff pair that will drive you crazy. The same applies to *dividers*.

3. *Pencils, Rubber, Torch*—Take your own; the last in case the ship's batteries are not as splendid as the owner says.

4. *Binoculars*—Take your own.

Generally speaking, remember that all unknown equipment is suspect. Put not your faith in electronics!—or at least until you have checked their accuracy. Make your own checklist for nav lights, compass light and all the other points that can go wrong and check what you can before you sail. Echo sounder accuracy can be checked in harbour—log accuracy not. But don't endanger the ship by unsuspicious acceptance of—to you—untried and unknown equipment.

CHAPTER TWO

PLANNING
AND PREPARATION

A very great number of navigational mistakes and inaccuracies arise as a result of the conditions prevailing in a small boat. Bad illumination at night, sea-sickness, bad weather, confusion and, above all, tiredness, lead to carelessness—to that moment of inattention when you can copy three for two and make two and two add up to five. What can you do to minimise this danger? The answer is to do as much of the work as possible before you go on board, or at least before you set sail. You will be surprised how much can be done to ease the work during the trip by careful preparation.

Let us consider the planning possible for a passage of about 200 miles, either cross-Channel, or along a coast, a short race, or indeed almost any passage that is not trans-oceanic. The preparations are the same as those which I should make for a race. The main difference between racing and cruising navigation is that one takes more care with the former; and to my mind the beginner should try to be as accurate as possible, and only afterwards, with experience, omit what he has found to be unnecessary for cruising. Some friends of mine recently took a large yacht from Cowes to Le Havre and back with only the tidal stream atlas because they had left the charts at home; such a risk is not advisable even with years of experience! For this chapter I shall presume that you are going to St. Malo direct from Cowes. You have got your charts, they are up to date, and you have a spare evening at home to prepare the trip to the fullest extent possible. You have decided to sail on Saturday, 4 June, 1977.

ROUTE

Which is the shortest way? With your dividers measure off the distance roughly on the small-scale chart:

via Needles—Casquets	158 n.m.
via Needles—Alderney Race	144 n.m.
via Bembridge—Casquets	170 n.m.
via Bembridge—Alderney Race	159 n.m.

Considering this list you have two decisions to make. First, you have to decide whether to go out through the Needles channel or to go east-about. This depends on two considerations: tide and time of sailing. Will the tide suit you? High water (Dover) is at 1339 (B.S.T.) Saturday afternoon, or at 0204 Sunday morning. The tide turns at the Needles three-quarters of an hour before H.W. Dover, so if you can get started by midday, or shortly afterwards, you will go out through the Needles. If, however, you cannot get away until the evening, when the east-going tide is just starting, it may pay you to go out to the east, particularly if the wind is westerly and you are faced by a beat against wind and tide in the Needles channel.

The other consideration which might affect you, on which you would make a decision at the last moment, is that if there was a westerly gale, or near gale, blowing, it would be more comfortable, even with a fair tide, to avoid the very nasty sea kicked up by wind against tide through the Needles channel. Going out eastward you would benefit from the lee of the island and could hope that the wind would moderate and veer to give you a lift across Channel. This, of course, is seamanship rather than navigation, but the two are very closely allied.

Your second decision is whether to go outside everything, round the Casquets, or to take the short route between Alderney and France. Studying this route you will see written on the chart 'Race of Alderney'. This should immediately put you on your guard. Looking at the larger scale chart (2669) you will find 'heavy overfalls', 'occasionally breaks', the signs for overfalls and tide-rips, and in the middle a tidal reference ◈. At the bottom of the chart under the tidal stream information you will notice that at ◈ there is a mean spring tidal current of 5·6 knots. All this is enough to tell you that:

1. You should not try to go through against the tide.
2. You should not try to go through in bad weather as you will find very broken rough water indeed.
3. If the weather is fine and the tide right you will get a wonderful lift on your way.

Now when will the tide be right for going through the Race? Again under ◈, at the bottom of the chart, you will see the direction of the tidal streams given for that point. The tide starts to turn in your favour four hours after H.W. St. Helier and runs SSW from five hours after until two hours before. You should, therefore, start to go through five or six hours after H.W. and not later than three or four hours before. Now look up H.W. St. Helier in the almanac and you will find H.W. (B.S.T.) Saturday 0902, 2023; Sunday 0849, 2108; Monday 0934. Note down lightly on your chart the optimum times of arriving off the Race. Saturday 1430; Sunday 0200, 1430; Monday 0300. This will save your having to look it all up again, and, when you get there, you can check your timing at a glance.

If the weather is bad you will go outside the Casquets. Studying this course you will notice the following points:

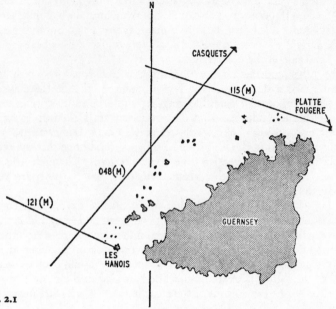

Fig. 2.1

1. You cannot go straight from the Casquets to Les Hanois; just NE of the latter there are some nasty unmarked rocks in the way, so you must work out a safety bearing. Draw a line from the Casquets which safely clears the rocks and draw two cross lines from Platte Fougère and Les Hanois; note on these lines the magnetic bearing from seaward of the various lights (Fig. 2.1). This means that from when the Platte Fougère light bears 115° until Les Hanois bears 121° the Casquets must not bear *less* than 048°. Having this already drawn on the chart will save you a lot of bother later on.

   If you cannot see the Casquets you must keep outside an imaginary point where Platte Fougère bears 094° and Les Hanois 202°, and be careful not to cut the corners, taking other safety bearings when necessary (Fig. 2.2).

2. You must not go closer to the Casquets light than half a mile to the west as there is a rock.

3. 'Violent eddies' are indicated off the Casquets so make a mental note to give it a good berth in rough weather.

4. You must leave the buoys of the NW and SW Minquiers to port, and if you are beating you must not go to the east of a line joining the two.

5. St. Malo has a very tricky entrance which you must study on the large scale chart (compare Cherbourg which you could get into using chart 2669, if need be).

6. You will see on the right of the chart, '*Tidal Information and Chart Datum*. St. Malo H.W. mean springs 39·7ft, L.W. mean springs 4·3ft', which means that at springs you will have a difference of about 36ft. If you anchor in 2½ fathoms of water at low tide, you will have 8½ fathoms under you at high water (or, to be up-to-date, anchoring at low tide in four metres you will have fifteen metres under you at high water). Make a note of this too, so that when you arrive you can tell the skipper how much chain he will want. Always check the range of tide at your port of arrival.

These are only examples to show how the chart, carefully studied beforehand, can warn you of dangers, tell you what to expect, and indeed give you almost all the information you need. For the rest of this chapter I shall presume that you have decided not to go through the Alderney Race.

Now, lay off the courses and measure the distances. Write out a list of these in the form of a table (see Table 2.1) and pin it up over the chart table.

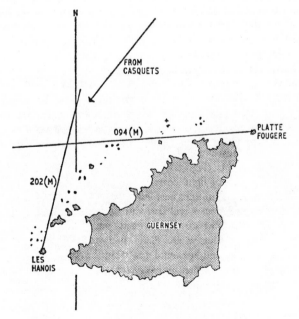

Fig. 2.2

Table 2.1

	Course (Magnetic)	Course to steer (corrected for compass deviation)	Distance (n.m.)
Needles–Casquets	218°	216°	66
Casquets–Les Hanois (clear of rocks)	228°	226°	21
Les Hanois–SW Minquiers ..	162°	159°	36
SW Minquiers–St. Malo ..	158°	155°	19

Personally I do not rule the courses on the chart, for the following reason: it is obvious that if you sail to a position not on your original course, this course is of no possible further interest to you: you have a new course to make—from your new position to your landfall. The original course, however, if drawn on the chart, acts in some strange way as a magnet to many people. It seems to represent some concrete highway, a highway which will serve them better than any other strip of water. Fig. 2.3 illustrates this. You are going from *A* to *B*, but for various reasons you find yourself at *C*. Your new direct course is naturally *CB*, and *AB* has now ceased to be of the slightest interest. The wind is SSE, so that it is clear that the starboard tack is the nearer and better tack, but you would be amazed how many people will wish to go off on the port tack in order 'to get back on course quicker'. The only way in my view to avoid this incomprehensible but fairly common error is not to draw the course on the chart.

It is, however, well worth while to mark on the chart a rose from the point you are making for (Fig. 2.4). On each line write the course (*M*), corrected also for compass deviation if necessary. With this rose you can, when tacking, or when otherwise necessary, read off your new course at once. You will see also that even if your destination is hidden, by the chart being folded, the course can be read. In this

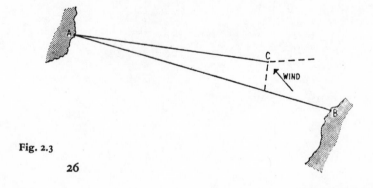

Fig. 2.3

particular case, the rose also automatically gives you a safety bearing past the rocks between Alderney and the Casquets.

LIGHTS

Pin up over the chart table a list of the characteristics of the principal lights you may see; this will help to identify them quickly and easily. Nominal range is now listed (the same as luminous for practical purposes), i.e. the maximum distance at which a light can be seen. However, if the range is 30M you are not going to see the light 30 miles away unless it is 100m high and your eye 20m above the sea. Conversion tables for arriving at the correct geographical range for your height of eye are in *Reed's* and the *Admiralty List of Lights*. As a rough guide, when your height of eye is 7ft (2m) you will see the Lizard light (70m) at 20 miles, and a light vessel (12m) at 10 miles. You will see any light with an elevation of 4m at 7 miles provided it is strong enough. Powerful lights with big ranges may well be seen much further away than expected either because a wave lifts you or because the loom is visible. In good conditions I have seen the loom of Cherbourg and St. Catherine's simultaneously.

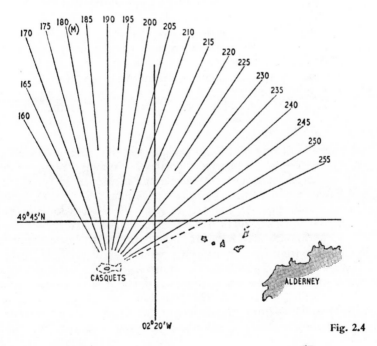

Fig. 2.4

Light	Characteristics	Range (n.m.)
Cherbourg (Fort, de l'Ouest)	Gp. fl(3) W.R. 15 sec	23 (19 red sector)
Cap de la Hague	Fl. 5 sec	24
Quenard Point	Gp. fl. (4) 15 sec	28
Casquets	Gp. fl. (5) 30 sec	28
Platte Fougère	Fl. 10 sec	16
Groznez Point	Gp. fl.(2) W.R. 15 sec	19 (17 red sector)
La Corbière	Iso. W.R. 10 sec	18 (16 red sector)
Roches Douvres	Fl. 5 sec	29
NW Minquiers	Gp. fl. (2) 10 sec	buoy
SW Minquiers	Gp. fl. (4) 15 sec	buoy
Cap Frehel	Gp. fl. (2) 10 sec	29
Les Courtis	Gp. fl. (3) 12 sec	8
Le Grande Jardin	Gp. fl. (2) R. 10 sec	15

Having made this list, check that no two lights have the same characteristics; occasionally this does happen. It may seem an absurd exaggeration to mistake two lights, say, twenty miles apart, but once, after a gale, a very good navigator mistook one light for another with the same characteristics, because the watch on deck, in order to keep dry, had borne away and sailed 40° off course for the best part of the night and had never mentioned it. A friend tells me that the need to identify lights carefully was brought home forcefully to an American in a U.S. Admiral's Cup boat in the 1973 Fastnet. He did not believe in the west-setting tide, managed to mistake Fowey for Looe, headed too far inshore, and threw away a cast-iron win.

Study the lights for coming into harbour (this is discussed in more detail on p. 73); the entrance to a new port can be confusing and the more idea you have got of how the lights are organised the better.

TIDES

For most Channel or North Sea passages, if you expect to be out for more than twenty-four hours it is worth while noting the times of the tides at the reference port on the relevant tidal atlas. H.W. Dover (B.S.T.) for 1st, 2nd and 3rd June, 1977: Wednesday, 1st 1114, 2336; Thursday 2nd 1203, 0025; Friday 3rd 1250. Write these down on the page for H.W. Dover and then, working backwards and forwards, note down the time for each hour before and after H.W. Dover. On the inside of the front cover write down the daily mean range at Dover: Wednesday 6.2m, Thursday 6.4m, Friday 6.5m. This will enable you at any moment to see what the tide is doing at any particular spot just by opening the atlas. At the same time have a good look at the atlas. Does the tide turn first inshore or out? Does the tide run harder inshore or offshore? Are there any points with a very fast tidal stream which must be avoided if the tide is foul? Are there any places where the tidal stream is so fast that you feel you must have further information before you go close (e.g. Pentland Firth, 10 knots; Portland Race, 7 knots)? The *daily mean range* at Dover allows precise forecasts to be made of the speed of the tidal stream. In some places there is only half a knot difference between springs and neaps; but 4 hours before H.W. Dover in the Alderney Race the stream runs at 4.0 knots neaps and 9.7 knots springs, a difference of 5.7 knots! The tidal atlas contains a table which, used in conjunction with the mean range at Dover, will give the correct speed for the day.

RADIO BEACONS

Round the coasts of Great Britain and most particularly in the Channel there are groups of four or five radio beacons; each group is on one frequency, the beacons transmitting one after the other. They are so arranged that in any area one frequency will give enough bearings for a fix. For example the Central Channel Group consists of six beacons (Portland, St. Catherines, C. d'Antifer, Le Havre, Pte. de Ver and Barfleur) which virtually surround a large area of the Channel. They all transmit on 291.9 kHz, each station transmits for 1 minute, the cycle for the group being 6 minutes. Thus Portland transmits at the hour, 6 minutes past, 12, 18, etc., Le Havre at 3 minutes, 9, 15, etc. Coloured charts of the beacons, their frequencies, call signs, range, times of transmission, etc., are available and should be kept handy.

SEA TRAFFIC SEPARATION ROUTES

Traffic is now so heavy that shipping lanes are recommended in many places, especially in the Channel. Yachtsmen are advised to keep clear of these lanes and to cross them as much as possible at right-angles. The general positions of these lanes are shown in *Reed's* and details are found on up-to-date charts. As far as our cruise is concerned there are lanes off the Casquets.

You have now done everything to ensure that your work aboard is going to be as easy as possible. You have, as far as you can, eliminated the need to look at the books. You have collected all the things you need, instruments, pencils, rubbers, pencil-sharpener (a knife is likely to be 'borrowed'), etc. You have laid off your courses and measured your distances, made a list of lights for easy reference, noted the times of the various tides, studied the chart, laid off safety bearings for any dangerous points; and finally you have checked your figures. You need only the latest forecast and a visit to the harbourmaster's office to glance at the latest *Notices to Mariners* and you are ready to start.

CHAPTER THREE
AT SEA

Once upon a time the owner of a yacht took some friends out for a day's sail. The distant coast and the hills behind gleamed in the sunshine, the boat slipped easily through the calm water, and at lunch time the harbour lay, hardly visible, miles away. With two gins and a good lunch inside him the owner felt he had earned a nap. When he woke an hour or so later, the wind was just too strong for full canvas. More anxious for his sails than his position, our hero lowered the genoa and set a small jib. Jib set, it was blowing harder still and he decided to reef. He had not done this for ages, the reef pennants were not rove, and with an inexperienced crew it took a longish time. When the job was finally finished the owner, now wet and cold, looked up to find the visibility closed in, no sight of land, and every prospect of the weather worsening.

Pulling himself together he went down to look at the chart (which luckily he had bothered to bring): 'We sailed more or less south west for perhaps three hours, at what speed? Say five to six knots; then . . . I wonder what the others did while I was sleeping? We must have been shifting but I wonder what the course was? What's the tide doing? I wish I had bought those new wireless batteries yesterday . . . where the devil are we?' Our hero had now landed himself with an unpleasant decision: should he work his way inshore, not knowing his position, towards a rocky coast, with a rising sea and heavy rain reducing visibility to five hundred yards, or should he wait, hove-to, till the weather cleared?

This fable is not at all impossible, *mutatis mutandis* it has happened many times and will happen again. The moral of the tale is that you must navigate *always*. Perhaps motoring has a bad effect: in a car first you get lost, then you look at the map, go back or ask the way. This is impossible at sea. A clear idea of the ship's position at all times is essential. The closer you are to land, the more accurately you must know it: in Beaulieu river you should know it to within 20 yards, in the

Solent within 200, near the coast within half a mile, in the Channel within five miles, in the middle of the Atlantic within, say, fifty. The navigator *must* mark his position on the chart at least every hour (except in the middle of the Atlantic). Make this a rule and stick to it.

Now, how can the navigator find out his position to mark on the chart? There are two ways: by dead reckoning and by 'lines of position'.

DEAD RECKONING

Dead reckoning is the name given to the method of plotting the position of a ship by means of distance sailed and course steered: that is by log and compass only. If you cross a familiar room, in the dark, and arrive without touching or hitting anything, you have achieved this by D.R.; by going so many paces in a certain direction, turning slightly left to avoid the table and right again to reach the door. If the light is turned on, you are only a few inches from where you thought you were. The room is only charted in your mind, but you have arrived, possibly without realising it, by plotting your course and distance.

At sea there is a chart which shows the position of the objects in the room, the log which measures your paces and the compass which tells you the direction. If you do not use D.R. you are like a stranger in the dark room, completely lost.

Let us look for a moment at what happens if you sail with no D.R. at all, and let us carry it to absurdity. You leave the Scilly Isles and sail for twenty-four hours. You use neither compass nor log. At the end of that day it is cloudy and your wireless is not working, where are you? If the maximum speed of the boat is seven knots, you may be anywhere in the circle in Fig. 3.1 that is not land or in sight of land! On the other hand, had you used D.R. (with no errors), you would know you were at A, 150 miles from the Scillies (log) on a course of 270° (compass). The object of dead reckoning is to reduce this enormous area in which you might be, this 'area of uncertainty' to as small an area as possible. This presupposes a known *point of departure*. It is useless knowing you have sailed so far in such and such a direction if you don't know where you started from! So when you are leaving a harbour or a narrow estuary for the open seas mark accurately the point of departure with the time and log reading (or stream the log).

If the sea were like a pool, without winds, tides or currents, and if every helmsman kept the exact course and if the compass had no deviation and the log were 100 per cent accurate, D.R. would suffice to take you round the world with no other form of navigation at all. The course for the Casquets from the Needles is 218° (M) and the

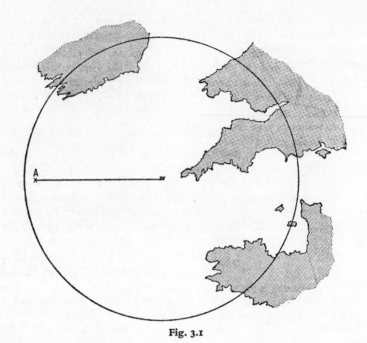

Fig. 3.1

distance 66 miles; in our mill pool, therefore, if you stream the log at the Needles at 0 and steer 218° you will hit the Casquets when the log reads 66. This just does not happen as we all know, or happens only once in a lifetime (as a child, I hit an unlit buoy at night, in the Thames Estuary, just as my father was saying we should be getting to it). What forces affect your D.R., what errors creep in and what then can you do to correct for them?

TIDES

The first and most obvious force which puts you off course is the tidal stream. Take the most simple example: if you go south for one hour at five knots and there is a one-knot easterly tidal stream you will have the plot shown in Fig. 3.2. You start from *A*, sail to *B* and the tide has set you to *C*. The next hour you start your plot from *C*, your corrected D.R. position. Consider the tides in the Channel leaving the Needles one hour after H.W. Dover with a mean range of 3.4m, course 218° (M). You have already noted the times in your tidal atlas, together with the mean range for the day at Dover. You do six knots

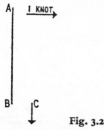

Fig. 3.2

on course, so during the first four hours (1, 2, 3 and 4 hours after H.W. Dover) you have the following tidal stream:

Knots	Direction (M)
1·6	270°
2·4	270°
2·8	270°
2·2	270°
9·0 miles	270°

Fig. 3.3 shows two methods of plotting this, (a) marking the tide every hour, (b) correcting for all four hours together. Which is the better system depends on the circumstances. It is more accurate to correct all four hours together, because you are not confused with fractions of an hour or such-like; thus at the beginning of a long trip it is preferable to correct for several hours at once, since in open waters it is the ultimate accuracy that counts, not the minute-to-minute position. Alternatively, near land, or approaching land, it would be better to correct every hour because the up-to-the-minute position is important. If your tidal stream is constant neither in direction nor in speed, then it may be easier to plot it hour by hour. Compare:

Hours	Direction	Speed		Hours	Direction	Speed
1	230°	2		1	230°	1·5
2	250°	2	with	2	250°	2
3	260°	2		3	260°	3
4	270°	2		4	270°	3
	mean 252°	8 miles			mean 256°	9·5 miles

Taking the mean direction of the first with constant speed is easy, taking the mean of the second needs more care and it will depend on you, individually, whether you wish to work it out or plot it hour by hour. Much of this work will be simplified by a pocket calculator.

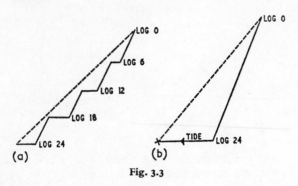

Fig. 3.3

The other question that arises from Fig. 3.3 is whether you should have allowed for the tide and steered more easterly, thus keeping along your direct course instead of being swept off it. This is really a question of strategy, but when you are crossing two tides, the flood and the ebb, as when crossing the Channel, it is better to ignore them and let them cancel each other out, rather than allow for each as you go along. You sail further if you keep on the direct course by changing course-to-steer to allow for the tide, than you do being swept away and back again. If you have only one tide to cross, however, then you must correct beforehand, and give the helmsman a course to steer which will keep him on his direct course.

Tidal charts are not very easy to use accurately, and great care must be taken interpreting them. At 1 hour after H.W. Dover, in the Admiralty *English Channel* atlas, immediately south of St. Catherine's Point there is an arrow marked 1·3, 2·4. In the Admiralty *Solent and Adjacent Waters* pocket atlas you will see, in what appears to be the same place, an arrow marked 1·7, 3·4 for 1¼ hours after H.W. Dover. Here there seems to be a discrepancy; but if you take your dividers and measure the distances you find that the first arrow is five miles from the Point while the second is only one mile.

Use the most detailed atlas available. Look on each chart to see if there are tidal-stream positions for which figures are not given in the atlas (for example there is a position off the Manacles on the chart which is not given in the atlas). Remember that usually:

1. Tides run more weakly in bays and more strongly off headlands.
2. Tides run more strongly near the coast and more weakly offshore.
3. Tides turn first inshore.
4. Tides set into bays.

5. Tides run more strongly in or near straits, e.g. at Dover and between Cherbourg and St. Catherine's.

Mark your course sailed on the atlas carefully so that you know where you have arrived, and if there are no figures just there take the mean between the two nearest sets of figures.

A word of warning is in place here. However accurate you are the tidal streams may, occasionally, have a nasty one up their sleeve for you. Michael Reeve-Fowkes, the compiler of *Stanford's Tidal Atlases*, says in his preface: 'Tidal predictions are no more, no less than that which they are named. . . . Many factors, emphemeral and incapable of being calculated in most cases, occasionally interfere with them, both in the case of heights, and in strengths and directions of flow of tidal streams.' Elsewhere he quotes the French *Courants de Marée* '. . . in good weather, and for streams of a maximum of 2 to 3 knots, the predicted speed at springs presents an approximation of 15% to 20% and the direction a possible error of up to 20°. At neaps, the relative errors on the speed are greater, but the streams being slower, the absolute error is comparable to that of springs.' This is not as bad as it sounds (it means little more than ½ knot at maximum), but you cannot accept the figures blindly. If you are in shallow waters use your lead or echo sounder to check frequently. Even today a prudent captain may prefer not to trust his charts and other information and Admiral Bini, who writes the Appendix in this book, took his ship up the River Plate preceded by a launch with a yachtsman's echo sounder to make doubly sure that his precious underwater equipment was in no danger of being scraped against the bottom.

Other than these notes it is difficult to give advice. How you will use the atlases and tidal information on the charts varies so much from place to place, with speed, direction, etc., that like all else in navigation the only answer is to go on trying until you gain experience.

STEADY BEARINGS

You can sail straight for a mark (buoy, entrance to harbour, etc.) and ignore tides, currents, leeways, the lot, *if* you can see land behind your objective. Making for a buoy across the Solent for instance note the tree behind the buoy and keep the two in line. The bow of the ship may be pointing 20° to port or starboard but never mind, you are going towards the buoy in the shortest way and will be adjusting your course automatically for changes of strength in the tide. This keeps the buoy on a steady bearing. If you are sailing into a stronger tide and are already nearly hard on the wind you may decide not to *head* for the buoy, but this is another matter.

Steady bearings work in reverse too—if you see a ship moving against the background she will *not* hit you, but you are on a collision course if she appears stationary against the background. Many who unconsciously and automatically apply this principle to their everyday driving and walking have trouble applying it at sea.

CURRENTS

Horizontal movements of water not caused by tides are called currents. In British waters these can be ignored; elsewhere, as in the Atlantic where the Gulf Stream flows, currents can be important. The tidal atlas has a paragraph on currents in the Channel which should be read, but no account need be taken of them for the purpose of D.R.

LEEWAY

Leeway is that component of the wind force which pushes a boat directly down wind. Leeway is usually measured as an angle. A good modern ocean-racer has negligible leeway, and any that does occur will be only when beating to windward. A big sea and a light wind provide the occasions when there is most likely to be leeway. It is difficult to judge and varies from boat to boat. In some motor-sailers or in a very shallow draft boat, I should allow two or three degrees of leeway, the bigger the sea the bigger the leeway. In a modern, deep draught yacht I should ignore leeway unless it was very rough. An error of 1° will make an error of about 1 mile in 60. When you think that if you misread your tides your D.R. may be out one mile in five you will appreciate that leeway is not of great importance.

When beating close inshore note carefully if there is leeway (this is particularly clear if the tide is slack) and you will soon get to know your ship well enough to judge for yourself when to correct for leeway and when not. Motor boats have much more leeway than sailing boats because of their shallow draught. When under power it is often possible to assess leeway by looking at the angle made by the wake with the fore-and-aft line of the ship.

COMPASS ERRORS

Parallax   When a boat has tiller steering there is always the danger that the helmsman may not allow enough for parallax. This means that from where he sits the lubber line appears to be lined up with (say) east on the compass card when the ship's head is really 095° or 085°. Sometimes this difficulty can be avoided by steering on the athwartships lubber line, but the navigator should try this for himself, warn the helmsman and check that he is not introducing a constant error into his D.R. from this source.

Magnetic Variation and Deviation   These have been discussed earlier but to recapitulate: the hand bearing-compass is taken to have no deviation (do not hold it close to shrouds or other steel) so if you are using the magnetic rose on the chart, and this is up to date, there are no corrections to be made to bearings taken with it. The steering compass has its own deviation card and any corrections given on this must be applied both to the course given you by the helmsman and to those given him by you.

As dead reckoning is so completely dependent on the compass it is clear that no possibility of additional error can be tolerated. If you suspect that your compass has some error that is not included on the deviation card it must be checked or re-swung. It should be noted that in big steel ships, the compass is usually checked twice a day. This, of course, is not necessary in a wooden yacht, but I would certainly check before a long passage or an important race.

LOG ERROR

If you use the length of line recommended by the maker you will not have a large log error. With a big following sea, however, logs tend to over-read, owing to the spinner going up and down hill through the waves. Illingworth gives 6 per cent as a maximum figure; I remember one occasion when, according to my reckoning, with a very big following sea indeed it was as much as 10 per cent. Fortunately this is a safe error; it does not lure you on to the shore. Too short a log line causes under-reading at speed as the spinner comes out of the water.

Whenever you have a straight passage, without changes of course, check the log against the actual distance run and you will gain experience of the conditions in which it over-reads and to what extent. It should be remembered, in spite of this warning, that normally the log is extremely accurate as long as it is turning freely.

HELMSMAN'S ERROR

This is the most variable and least discussed of errors, but to my mind one of the most important. Since the navigator cannot always steer the boat himself he must largely rely for his D.R. on what the helmsman says he has steered. The accuracy of D.R., therefore, is closely bound up with, and dependent on, the reliability and ability of the helmsman; both his ability to steer the boat and his ability to know what he has steered. From the point of view of D.R. it does not matter the helmsman steering 160° instead of 180° as ordered (he may have a good reason), but it is essential that the navigator should *know* that he has steered 160°. This means that the navigator must know his helmsmen. He should know who gets the best out of the ship beating in light winds; who takes her well with a heavy following sea; who is experienced and who a beginner; who concentrates and who could not care less. Equally, the navigator should know his boat. Has she a weather helm? Is she unruly with a following sea? You may say that all this is the skipper's job, that it has nothing to do with navigation and that anyway if you are not racing it does not matter, but whereas when racing, everyone is on his toes, crews get more careless and slack when cruising. This applies to all aspects of navigation. More 'incidents' occur cruising and coming back from ocean-racing than during a race itself.

A long time ago we sailed back from France, Lézardrieux, I think, to the Solent. We made the crossing at night on the port gybe with a rising south-wester dead astern and a biggish sea. The owner, an excellent navigator, had a bad sinus infection and had to stay below, relying on us to pass down to him log-readings, courses and any other information. We were all inexperienced (though we thought we knew the lot), with little knowledge of the boat and none of us realised that, from fear of gybing, we were heading very much higher than the course we had been given, and which we believed ourselves to be steering. Our course must have been 040° and we steered about 020° It seems impossible looking back on it now. I cannot remember whether we saw any lights or if we succeeded in misreading these too (wishful thinking can be very dangerous), but I do remember, as dawn broke, a cry from deck as, through the slight mist, the watch saw the high hills between St. Albans and Weymouth close in front of us. We had narrowly escaped going straight into the Portland Race and were thirty miles off course. Obviously, had the owner been able to come up on deck this would not have happened, he would have taken the helm himself, seen that the course was too difficult to hold, given us a more weatherly one, plotted his D.R. accordingly and gybed when necessary. The other safety factor would have been a tell-tale compass below, which he could have watched and seen that the mean course was far removed from that which we were reporting to him.

Any navigator should try to steer any boat he is navigating in various conditions to see if she has any difficult points of sailing; if he does not know the crew he should watch the compass himself, frequently and for long enough periods to make sure the course is the same as that reported by the helmsman. Try to watch the compass inconspicuously, so that whoever is steering does not realise you are checking up; he will be more careful than normal if he knows he is being watched. An experienced helmsman will of his own accord put any difficulties to you or to the skipper; 'she's very difficult to hold on this course', 'I must either ease sheets or head her up a bit', or something of this sort. I should feel more inclined to watch someone who, in difficult conditions, reports that he has steered the course ordered with no comment; it may show that he is not even aware that there may be something to discuss. But this is a psychological problem which everyone will solve in his own way. The main sources of trouble are to be found in the situations described below.

Beating   In a fresh gusty breeze the helmsman thinks he is laying the course but with every gust heads up (and speeds up) five, ten or even fifteen degrees. When asked his course he gives it for the period when the wind is lighter, without taking the gusts into account. Fig. 3.5 shows what he is doing to your D.R. Although beating well up to windward he is reporting a course seven degrees off; and seven degrees

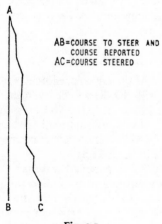

Fig. 3.5

means seven miles out after sixty miles: just a fast night's sail. At night a helmsman may ramp off to get the sensation of moving, or even to get less wet. This last is understandable, but unforgivable if it entails deliberately giving you the wrong course.

In a very light breeze the boat may tend to have lee helm, and in order to keep her moving the helmsman may steer more to leeward than he realises. This is less nuisance to the navigator, as with a light breeze the distance covered is much less.

Reaching   With a fresh breeze and a quartering sea, if the sheets are not trimmed right (and even if they are sometimes) the boat may carry considerable weather helm and the helmsman tend to be always to windward. Perhaps of all courses, this is the most difficult to judge the mean course steered, even with experienced helmsmen. I know of one boat, racing down from the Fastnet to the Bishop Rock in a south-west gale, which found herself many miles to the westward because she was pulling up to windward all through the night without the crew realising it. She won her class in spite of losing a good bit of time, which shows that even in that class of racing these errors may not be easy to recognise and allow for.

Running   As illustrated by the story I told earlier there is always the danger that when running without a spinnaker, if you give a course too down wind, the helmsman will head up continuously in an effort to avoid gybing. He should be given a more weatherly course and a strong preventer. The opposite may happen on a shy spinnaker reach as he may bear off in gusts to prevent the spinnaker collapsing and the subsequent broach.

'Speedometer error'   If there is a speedometer in front of the helmsman there is always a risk that he will be hypnotised by it into sailing a faster course, but not the one ordered. To gain speed he may bear away when on the wind, or head up when broad reaching or running. As his eyes are glued to the speedometer his course as reported to the navigator may leave something to be desired.

No rules can be laid down about helmsman's error, it depends on him, on you and on your judgments. It may well be that you sail for months without meeting that particular combination of boat, sea, wind and helmsman which will give you a serious error, but do not forget to check up, or one day you will have a nasty shock.

PROPELLER ERROR

In some boats with only one propeller there is a strong steady pull to one side or the other under power. If the boat tends to go to starboard for example, the navigator should give a course two or three degrees to port of that required so that the almost inevitable error of the helmsman is compensated for.

D.R. NOT RECORDED

The last of the errors that may creep into your D.R. is one of the commonest: inability to plot the D.R. position accurately because no one has written down the information. An hour goes like this (with no entry in the log):

1200–1225	course steered 250° 2 miles
1225–1240	wind freshens and veers, course 270° 2 miles
1240–1300	wind veers further, course 290° 2·5 miles

Now unless your helmsman has a very remarkable memory he cannot remember the times and distances and may give you: 'Well, 250° at first, then almost at once I was headed off and have been doing about 280–290° since'. So you must lay off six and a half miles as best you can. The two variants for the hour are shown in Fig. 3.6.

Fig. 3.6. AB = course sailed; AC = course plotted from insufficient data; BC = 0.9 n.m.

The log should be written up at least every hour, preferably every half-hour and when racing every quarter of an hour. Try to make it an automatic habit with your crew to note down changes of course, sail changes, wind changes, etc., with the time and log reading. A sample log is given in Table 3.1.

Table 3.1

Time	Log	Course ordered	Course steered	Wind	Barometer	Remarks
1200	23	90	90	WSW 3	30·2	
1230	25	90	95	W by S 3	30·2	Wind tending to veer
1300	28	100	100	W by S 3½	30·1	St. Catherine's bearing 355°. Set small spinnaker
1310	28·5	100	100	W 4		Gybed; wind tending to veer and freshen.
1330	30·5	80	80	W 4	30·0	

RELIABILITY OF THE D.R. PLOT

In what follows D.R. is used, as it so often is in practice, for the estimated position. That is to say course and distance run, adjusted for tides, currents, leeway and any of the errors mentioned in the last few pages.

It has been said that to plot your position is easy, to plot it accurately is difficult, and that to estimate its accuracy is very difficult indeed. We have seen why it is difficult to plot the D.R. position accurately because of all the various errors which may creep in; now, when you have your plot on the chart, how much reliability can you put on it? How large and what shape is the 'area of uncertainty' round the plot? As you approach a coast in poor visibility, how are you going to answer the question: at the worst where are we?

The first thing to remember is that the size of the error in the D.R. plot increases with the passage of time as well as with distance. The area of uncertainty after 100 miles will be larger than after 10, it will be greater after ten hours than after one. It will also be larger if you do 100 miles in twenty-four hours than if you do the same distance in twelve hours. Fig. 3.7(a) presumes an equal possibility of error in all directions. If, however, you have been reaching all the way, fairly fast with a steady breeze, your log is quite reliable but perhaps you think that the steering, for various reasons, has been inaccurate; then your area of uncertainty would look like that shown in Fig. 3.7(b). Supposing on the other hand that your headings had been good but you are in waters where there is sometimes a north or south current, and you are uncertain as to the distance covered, Fig. 3.7(c) might be your area of uncertainty. I would say, however, that shapes as in (c) are rare, while

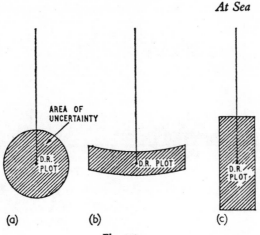

Fig. 3.7

(*b*) can only occur on a straight passage with no tacking. As soon as you start to tack the area very soon becomes polygonal and finally round, and only in exceptional circumstances would I draw an area other than round.

It will be noticed that the D.R. plot is in the middle of the area of uncertainty and this must be so. Your argument may go like this: 'log reading for the 4-hour watch, 28 miles; I know the log over-reads at least 5 per cent with a big following sea, corrected log reading 26·5'; or 'I gave a course of 180°, but Jack, who steered about a quarter of the time, had a mean heading of at least 195° and so probably did John, although I was asleep at the time, so I'll take a mean for the whole watch of 187°. This is what should happen, and does with experienced navigators, but I think many beginners plot what is written down and then at the end, when hoping to make landfall, think: 'this is my D.R. plot (or area of uncertainty) but I am sure I am more to the east'. If you think this you must work out this sense of being more to the east, logically, and reposition your D.R. plot so that it is in the centre of the area. If in your mind it looks like one of the shapes in Fig. 3.8, then your D.R. plot *must* be moved until the area of uncertainty is a circle with the D.R. plot in the middle.

Here the word *logically* must be stressed. There are occasions (fairly common in the air) when a navigator gets an overpowering sensation of being farther (say) north; this feeling has no concrete foundations and is not, in fact, true. If you get such a feeling try to find a reason for it; but if there is none you must get it out of your mind.

I think these sensations tend to come when approaching some danger. If, when in a fog, land lies to the south you may get a strong feeling that you are too far south, if to the west you are sure you are

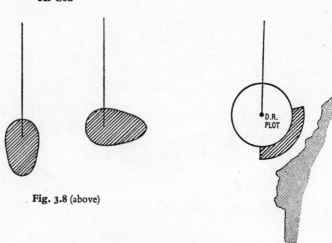

Fig. 3.8 (above)

Fig. 3.9 (right)

too far west. This is quite understandable, and you can, of course, put a safety margin on the dangerous side of your area of uncertainty (Fig. 3.9).

No navigator, after making a landfall ten miles farther south than he meant to, should be heard to say 'yes, well I did think we were probably a bit more to the south'; that thought, if based on valid argument, should have been incorporated in his D.R. plot. I have stressed this point because I am sure it is one of the main sources of error in amateur navigating.

The reader can now very justly complain that I have given no size for my original area of uncertainty; the circle to be used after a fastish, fine-weather, nothing-out-of-the-ordinary, mixed beat and reach in waters free from strong tidal streams. Very tentatively I would suggest taking a radius from your D.R. plot of 5 per cent of the distance sailed. In fog, however, I should enlarge this to 8 per cent for safety's sake. As I say, these figures are tentative. Each time you make a landfall check how far from your D.R. plot you really are and work out the area of uncertainty so that you can build up your own data on which to base your assumptions.

It is most important to find out why your position was wrong. Admiral Wyatt, Hydrographer of the Navy, when he navigated *Bloodhound*, would re-work his figures if he was even a mile out after a cross-Channel run, in order to discover the reason for the discrepancy. He was an expert and the amateur will probably be more than satisfied if he is only a mile out; but if bigger discrepancies occur (and they will) turn detective and hunt for the reasons. Not only the D.R. must be checked; one of the instruments discussed below may have been wrong, or too much reliance put on a faulty bearing or such like, but a reason there must be and if you can find it you can probably eliminate it in the future.

To summarise then, dead reckoning is essential and must be continuous, since it is the only system which can never fail you. It is liable to be incorrect owing to errors from the following sources:

1. Tides.
2. Currents.
3. Compass parallax.
4. Log reading.
5. Helmsman's steering.
6. Leeway.
7. D.R. not recorded properly.

A careful estimation of all these will give you the most accurate D.R. possible, but, *never* miss a chance of checking your dead reckoning.

LINES OF POSITION

For the purposes of this book by 'lines of position' I mean any line, position line, bearing, even depth contour, which can assist the navigator in establishing his position. These lines of position are used to check and correct your D.R. position and should be made use of whenever available. John Illingworth tells the following story in *Offshore* when stressing the importance of checking the position of D.R. plot as the case may be:

'... the first mark was out of sight of shore. We started taking back bearings shortly after the start, and found we were being set much more than the *Tidal Atlas* suggested was to be expected. This enabled us to sharpen up into the wind and when the mark came in sight, we were just enough up tide to weather the mark, though the wind went light. The other forty boats did not make it, and had a weary beat up against the tide. So we won that race too. We were paid some nice compliments about our cunning tactics, but of course, it was not a matter of cunning but just a little care in plotting'.

Or rather, I would have said, in checking their plotting.

Let me give you another example, taken from an account in *Yachting World* of February 1972, of the loss of a yacht at 3 a.m. off the north coast of France, the writer after describing the shipwreck continues:

'There was no excuse at all for running aground at this place and time. We had had two lighthouses in full view for at least two hours, and good radio fixes from Roches Douvres for more than that. One can only surmise at the probable reasons. The first is that the $1\frac{1}{2}$ knot foul tide

45

(which was still running in the offing), for which we had been allowing for three hours, failed in the "shadow" of the Les Triagoz plateau, thus giving us our full five knots over the ground in that last hour—we were one and a half miles further on than we thought.

The second reason was that it was 3 o'clock in the morning, at which hour no man is at his best; we were so certain that we knew where we were, that we probably grew careless in our plotting or steering or both.'

(Readers might like to know that all ended well and the crew picked up after some hours in their life raft.)

Now how do you check your D.R. position? By lines of position which can be obtained by:

1. Visual compass bearing.
2. Celestial observation.
3. Radio bearing.
4. Echo sounder (or lead).
5. Consol.

Let us look at a rather fantastic example just to see why I put all these together. In Fig. 3.10 you have checked your D.R. position by four methods, each of which gives you a line of position: a compass bearing from the lighthouse, a radio bearing from the radio beacon, a position line from the Sun and finally a check from the echo-sounder which puts you on the 6 fathom line. You could not be in much doubt, in this situation, in moving your D.R. plot to *P*. I have not included Consol because it is not to be used for landfalls.

You will see that each piece of information gives you a line and only a line (except for that from the echo-sounder which may give you an area instead). In this instance all the lines of position cross at the same point so no problem arises as to reliability; they are all equally good. But let us look at the same picture slightly changed. In Fig. 3.11 the radio bearing, the compass bearing and the Sun sight make a large cocked hat, a long way from the D.R. position, and all in a flat-bottomed bit of sea with a level sounding of six fathoms. Which piece of information are you going to trust? What is the relative reliability of each line of position? There is no rule to follow, each case calls for your judgment and experience.

Let us consider them one by one, presuming that you have had no outside information, that is to say have been on D.R. only, for the last twenty-four hours, through a nasty blow, and may be very lost indeed.

VISUAL BEARINGS

Make sure you are looking at the right mark (lighthouse, light vessel, etc.). By day, check the buildings with the description in the *Pilot*,

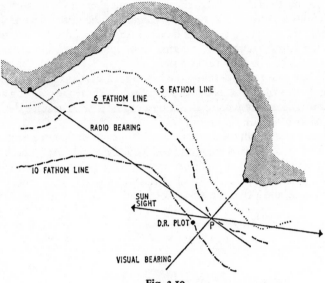

5 FATHOM LINE

6 FATHOM LINE

RADIO BEARING

IO FATHOM LINE

SUN
SIGHT

D.R. PLOT● P

VISUAL BEARING

Fig. 3.10

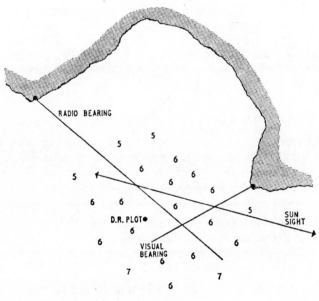

RADIO BEARING

5

5

6

6

6

5

6

6

6

5

D.R. PLOT●

6

6

VISUAL
BEARING

6

7

6

7

6

SUN
SIGHT

Fig. 3.11

light list or *Reed's*; at night, check the timing of the light with a stop watch and make sure that no nearby light has the same characteristics. Do not take a bearing of the land where 'the mark must be' unless you are very sure of yourself. Land seen from a different angle to the one you expect can be very misleading.

You must also consider how the bearing has been taken; with a hand bearing-compass by someone experienced; by someone glancing along the line of the steering compass; or by a beginner who steadied the hand bearing-compass on the shrouds?

I have included all the 'ifs' and 'buts' only to underline the variety of possibilities, the eternal question mark which must always be present in the navigator's mind. Apart from such queries, a compass bearing on a known and recognised mark is by far and away the most reliable line of position, and of course that most used, since it is the basis of all coastal navigation. One variation of the visual bearing should be mentioned here. Two known objects, both marked on the chart, when in line give a perfect bearing without the use of a compass.

CELESTIAL NAVIGATION

An experienced navigator with the necessary equipment on board, a clear horizon and a sight of the Sun, Moon or a star, will probably put a position line from one of the heavenly bodies next in reliability. Celestial navigation is neither difficult nor arduous, but it does need practice. If you are prepared to take the trouble it undoubtedly gives you more accurate position lines than any other method, except visual bearings, and allows you to be, to a considerable extent, independent of other aids.

RADIO BEACONS

The reliability of radio beacons, or rather of the bearings taken from them, has two aspects, the liability to error of the transmission itself and your own liability to error when taking the bearing. The transmissions from radio beacons become unreliable when:

1. The beacon is not within the listed range.
2. The beacon is along the coast from you.
3. The beacon is on the other side of hills.
4. The beacon does not give a complete null, that is if there is no position of absolute silence.

The transmissions are also less reliable at night and particularly so at dusk and dawn.

The possible sources of error from the ship are numerous:

1. A continuous metal lifeline all round the ship can cause serious deviations; this can be obviated by putting an insulator, which breaks the circle, somewhere along the lifeline.
2. If the boat is heeled over this may induce an error.
3. If the boat is swinging in a seaway it is only too easy to make an error in the compass reading.
4. The bearing is more reliable ahead or abeam than broad on the bow or on the quarter.
5. The bearing is ambiguous, i.e. if the beacon bears north it may equally bear south, and, when the beacon is on a light vessel, you may think you are heading for it whereas you are sailing away from it.

The *relative* reliability of beacons depends on their distance and listed range. If you are 50 miles from a beacon with a listed range of 100 miles, and 25 from one with a listed range of 50 the two bearings are equally reliable. It must be remembered, however, that an error by *you* in reading the bearing will be twice as big in the bearing from the beacon 100 miles away. It has been suggested, that after a run from the Fastnet in 1971, that large spinnakers can cause errors. I have no confirmation of this but bear it in mind that it could be so.

Take a series of bearings on a fine day, when you know your position from visual bearings, and do so whenever the opportunity offers so that gradually you can make up your own mind when and how much you can trust bearings from radio beacons, then, in fog, you will have some data to go on.

ECHO–SOUNDERS

Good sets can be very accurate; but in some places the bottom of the sea is flat, and while a reading of twenty fathoms may assure you that you are in no danger, it may not be much help in plotting your position. If, for instance, you approach St. Catherine's from the south, the soundings from ten to one mile off read: 18, 18, 19, 20, 18, 22, 15, 14, 25, 17, 8, which are not going to help much till the last reading. On more regularly sloping coasts, however, an echo–sounder can be of great assistance and when approaching land, in fog, it is invaluable as a safety control.

LEAD

The hand lead merely does, laboriously, what the echo–sounder does automatically and continuously. If, however, you have no echo–sounder or it is not working, the lead can be useful, and indeed essential when nearing land in bad visibility, tacking close inshore or coming into unknown harbours.

CONSOL

Consol is described as a long-range aid to navigation and warnings are given on the charts that it should not be used for making landfall. It is liable to great inaccuracies in certain areas and must be used with considerable caution. I would trust information obtained from any other source more than a position obtained from Consol. I would also be loath to correct my D.R. plot from it unless I had been some time at sea and was very doubtful of my position. However, with this caution it can be most useful. If you choose the right stations, the right time of day, and take enough counts, Consol can be extremely accurate. At any rate try it in good weather, see what results you get, and use it when necessary, not forgetting its limitations.

THE RELIABILITY OF FIXES

Two lines of position give you a fix; i.e. the second line tells you how far along the first you are. The nearer the cut is to a right-angle the less liable the fix is to error. Fig. 3.12 shows two fixes where the distances from *A* and *B* to the cut are equal. With the possibility of a 3° error on each side of the bearing it can be seen how much larger the area of error is when the angle-of-cut is very obtuse or acute. Whenever possible check your two-line fix with a third line. This will give

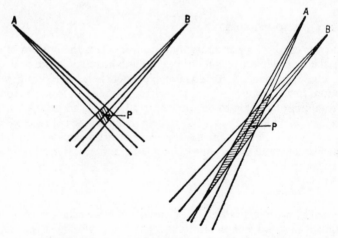

Fig. 3.12

you a cocked hat, as in Fig. 3.11, but practically never a perfect fix as in Fig. 3.10. Here again it is important for the cuts not to be too obtuse or acute. Fig. 3.13 shows this. When your cocked hat is small (Fig. 3.13, left) you can put yourself in the middle of it. This is not necessarily logical, but in practice it is the best you can do. If your cocked hat is larger you must consider the following points:

1. What are the distances of the marks or radio beacons you have used? An error of 1° in a radio bearing from a beacon, 60 miles away, is 1 mile; the same error in a visual bearing on an object 6 miles away is 200 yards. So the greater the distance the more error there is likely to be in the bearing.

2. In what circumstances were the bearings or position lines taken? In Fig. 3.14 line *A* is a position line from the Sun taken in excellent conditions, unaffected by a biggish sea; lines *B* and *C* are radio bearings, *B* 25 miles away, *C* 50 miles away, both difficult to take because of the violent swinging of the boat. You will probably prefer to put yourself at *X*, on *A*, which you trust, and nearer *B* than *C*.

3. Still considering Fig. 3.14, if all three were visual bearings,

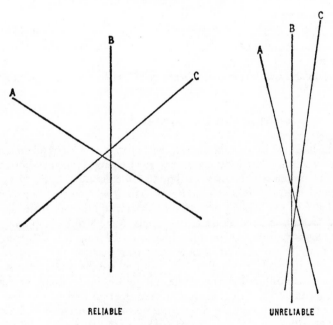

RELIABLE   UNRELIABLE

Fig. 3.13

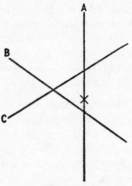

Fig. 3.14

A near, B some way off and C hardly visible, you would also put yourself at X.
4. Is your cocked hat too large? If your cocked hat is very large you should retake the bearings. 'Very large' cannot be defined too closely; for it depends on your distance from land; you could accept a cocked hat fifty miles offshore which would not do close inshore.

It must be remembered that a fix no more 'fixes' the position of the boat than does the D.R. plot. A fix has its own area of uncertainty too, and there will be occasions when you will prefer to trust your D.R. position and wait to get better bearings before correcting your position.

TRANSFERRING A LINE OF POSITION

In the fixes we have looked at, all the bearings have been taken almost simultaneously, but what happens if one bearing is taken an hour after the first, when the boat has sailed quite some distance? In Fig. 3.15 you have taken a Sun sight at 1500 hours. After two hours' sailing you see land and take a bearing of 200° on a distant lighthouse. According to your D.R. you have, in those two hours, done 8 miles on a course of 247°. Draw a line (*AB*) 8 miles long on a bearing of 247° from *any* point on the position line, then transfer your position line from *A* to *B*, drawing it parallel to the first. The point where the bearing crosses the transferred position line is your fix. Of course the reliability of this fix will depend on the accuracy not only of the two lines of position, but also of your D.R. during the two hours. I have included a Sun sight here because I want the reader to see how celestial navigation can be,

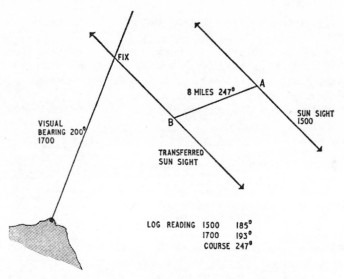

LOG READING 1500   185°
1700   193°
COURSE 247°

Fig. 3.15

and is, used in conjunction with other aids and is not something quite apart.

Fig. 3.16 shows another example of transferring a line of position. At 1120, log reading 63, lighthouse A (tower B invisible) bears 83° at 1230, log 70, course sailed 105°, tower B (A invisible) bears 360°. Transferring the bearing of A seven miles on course 105° gives you a fix.

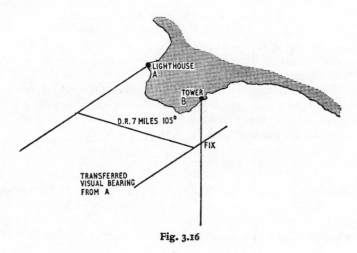

Fig. 3.16

53

DISTANCE OFF BY SEXTANT ANGLE

If you are near enough to a lighthouse to see both the sea-shore below it and the light clearly, the distance off can be measured with a sextant. Knowing the angle between the water at the foot of the light, and the height of the light from the almanac, chart or light list, the side *AB* of the triangle (Fig. 3.17) can be deduced. There is a table in the

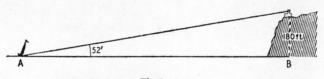

Fig. 3.17

almanac which, from the sextant angle and height of object in feet or metres, gives the distance off. This is particularly useful racing, when you can set a safety angle on the sextant (you must not go closer in than one mile and two cables, height of light 140 feet, sextant angle 1°06′) and when you see that you are getting too near, by the approaching coincidence of the light and the waterline seen through the sextant, you can edge further out. Since the height given in the almanac is that above high water, the lower the tide the further off shore this method keeps you.

SYNCHRONISED FOG SIGNALS

There is one other method of finding distance off. At Cumbrae and Cloch Point on the Clyde, and at Cherbourg, there is an ingenious system for use in fog. Just as you measure the distance of a thunder-storm by counting the seconds between the lightning and the thunder so the difference between the speed of radio waves and the speed of sound is used to measure the distance off the lighthouses. A series of 'pips' on the radio is synchronised with the acoustic fog signal. By counting the number of pips received before the signal is heard the distance off is measured; each pip represents 200 yd.

MAN OVERBOARD

These then are the main methods for checking your D.R. position. All of them need using with common sense in order to estimate their appropriate uses and relative reliability. There is, however, one particular case of fixing and D.R. which needs a little forethought:

a man overboard, if he is not picked up at once, means a very special job for the navigator. He, or someone for him if he is also the skipper must:

1. *Immediately* read the log, take the time and note the course. Even if he cannot get below to write this down he must remember it.

2. Keep an accurate large scale D.R. so that the boat can be sailed or motored back the way she has come, when she can be turned round to pick the man up. This may be very important when carrying a spinnaker at night. The D.R. is greatly simplified by not having to take the tide or current into account as they will affect ship and man alike.

3. *As soon as possible* position the boat as accurately as he can so as to reduce to a minimum the area for an eventual search by aircraft or ships.

Circumstances vary too much to say more, but every navigator should give this problem some thought so that he is prepared in an emergency.

Before leaving the subject of plotting and positioning I should like to give one final example of the importance of always knowing your position. A year or two ago we sailed from Naples to Ischia. Off the island of Procida, two miles from Procida harbour, five miles from the port of Ischia, with Cape Miseno a couple of miles behind us, the storm which we had seen coming up for the last half hour hit us. The *Tropea* is the typical summer storm of the Bay of Naples. It lasts half an hour, rises from a flat calm to over force seven in about one minute, turns the whole bay white and comes with such blinding rain that you cannot see the bow of the ship. With only a tiny jib we were doing six or seven knots, but because we knew exactly where we were, there was no danger. We laid off a course which took us out to sea and went safely and happily ahead until the visibility cleared and land reappeared. Had we been uncertain of our position we should have passed some very nasty moments. Only when you know where you are can you proceed with confidence to answer the question 'what shall we do next?'

CHAPTER FOUR
STRATEGY

With strategy we leave the 'where am I?' part of navigation for the moment and come to 'what shall I do next?' It is all too commonly considered that strategy belongs to ocean racing and that you need not bother about it when cruising. But I profoundly disagree with this. Strategy can save you hours of sailing; arriving in port in time for tea instead of having an extra three hour beat in the cold against the tide. Strategy can get you in before the blow comes on. Strategy can save you trouble and danger and loss of time. After all what is the difference between racing and cruising? Racing, you carry more sails; racing, you take calculated risks that you would not do cruising; cruising, you use the engine in light weather. Otherwise, whether you are cruising or racing, you want to get into port as soon as possible, provided you are not only out for a nice day's jolly. Indeed strategy may be more important when cruising than when racing. If you spend several unnecessary, cold, wet hours at sea when racing, you will lose the race but you have a full, probably experienced crew on board. Cruising, you may have wife and two children and a friend who suffers from sea-sickness and does not know anything about boats. It is evident that, in the second case, any situation is more difficult to deal with, more tiring, and, if you are caught by a blow, more dangerous.

What will influence your decision as to what course to steer, when to tack or gybe, and so on? The two main considerations are the weather and the tides.

WIND AND WEATHER

It is not my intention here to go into the science of meteorology. It is a very complicated subject which you must know profoundly before you can put it to good use. In British waters the weather is local and changeable and the forecasts are at their most reliable for strong winds or gales. If the winds are light you often find either that you get something different to what is forecast, or that you already have the forecast wind but are given no indication as to when or in which direction it will change. The weather is a continuity, however: what happens today depends on what happened yesterday; tomorrow's sun or rain on today's thus continuous forecasting is necessary for good results. On blank forecast sheets write down the shipping forecasts each day for three weeks (a tape recorder would help) and with the help of a book on weather draw the synoptic charts. Then compare yours with the charts in the newspaper and your own barometer, and look at the sky and note the weather. It is interesting that Alan Watts, writing on the meteorology of depressions in Adlard Coles' *Heavy Weather Sailing*, says: 'the barometer at sea is not a very reliable guide, because seeking shelter is a question of decision several hours ahead of the real blow. It must be the sky that demands attention.' The best most of us can do, however, is to listen to the forecasts and rely on what they say. How can you best use them if they are right, and guard against the possibility that they may be wrong?

BEATING

The object is to get from one point to another without wasting time; to do this when forced to beat you must snatch any advantage the wind gives you immediately, use all this advantage at once and then wait until a change of wind gives you another lift.

First consider beating towards a distant point dead into the wind. If the wind was steady, and you knew it was going to remain steady, you could perfectly well make two legs only, *ABP* in Fig. 4.1. But in northern Europe the wind seldom blows from exactly the same direction for a long time; indeed we all know the sensation that, whichever way you point the boat, the wind immediately comes dead

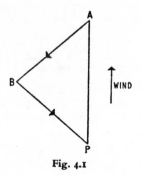

Fig. 4.1

on the bow. If, when at *B*, the wind backs, it has headed you again and you still have *P* dead in the eye of the wind; if, on the other hand, when near *B* the wind veers and frees you, you will have overshot P and sailed further than necessary. If *P* is to windward any wind change is certain to give you some benefit; if the wind changes enough, in either direction, it may even free you (Fig. 4.2). Now obviously you

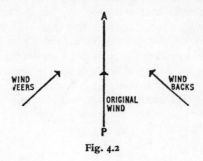

Fig. 4.2

cannot tack every ten minutes in order to keep *P* in the eye of the wind, hoping for a wind shift, so the best thing to do is to tack in a sector extended *downwind* from *P*. The recommended arc of this sector is 10° (see *Offshore* by Illingworth). Fig. 4.3 shows such a sector. It is wide enough to allow you to take reasonably long tacks but at the same time permits you to make the best of any eventual wind change. Of course,

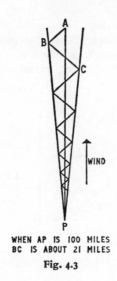

WHEN AP IS 100 MILES
BC IS ABOUT 21 MILES

Fig. 4.3

as you approach P, strengthening tides near the coast, or other considerations, may make it advisable to leave your sector. If you have marked your chart as shown in Fig. 2.4 the sector is ready for you to use, and you have only to keep your D.R. plot inside it.

If, when you leave A, the wind is not dead ahead then you must 'snatch your advantage', take the better tack until you are in the downwind sector from P and then carry on as before (Fig. 4.4). The

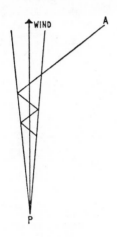

Fig. 4.4

better tack is that which takes you nearer to P. If tacking through 100° you can lay up to 30° from P on the port tack and up to 70° on the starboard, then the port tack is the better one.

Fig. 4.5 shows a more elaborate version with several wind changes. When you arrive at A (your D.R. plot) the wind heads you, so that you can no longer lay P. Your course beating to P is as shown, moving from the old to the new downwind sector as soon as the wind changes. It is the best way, with these winds, of getting to P, provided you have no idea whether the wind is going to veer or back.

Up till now we have presumed that there is an equal chance of the wind veering or backing. In Fig. 4.6(a) two boats, at A and B, are becalmed, equidistant from P. When the wind comes in from 205° which is in the better position? A can reach P with a tack at D, while B can reach P by tacking at C. In this case if A sails 10 miles to reach P, B sails 11·7 miles. If the wind comes in from 223° (Fig. 4.6(b)) A can just lay P, and where A sails 10 miles, B sails 13·5 miles. In both cases A and B were 3·1 miles apart at the start. Of course if the wind

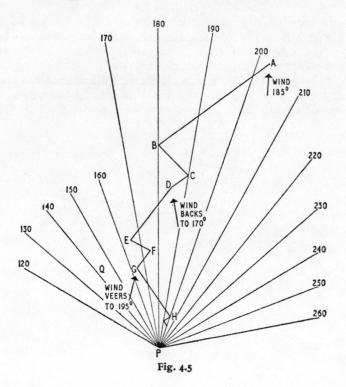

Fig. 4.5

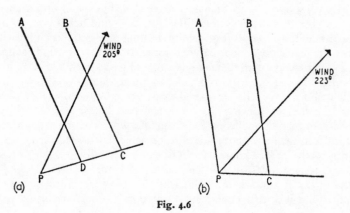

Fig. 4.6

came in more westerly still, it would free *B*, as well as *A*, and *A* would only have the slight advantage of being freer and therefore sailing faster.

These two examples show that the boat *nearer to where the wind is going to come from* has the advantage. Looking again at Fig. 4.5 you will see that if, after the wind had backed, you had had reason to think it would veer again, instead of doing tacks *EFG* you would have continued on the port tack to *Q*, *towards the coming wind*, and could have then laid *P* from *Q* when the wind veered.

If you expect the wind to veer, therefore, but do not know when, or how much, what should you do to be in position *A* and not in position *B* in Fig. 4.6? You should beat in the 10° sector (*AB* in Fig. 4.7)

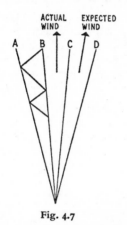

Fig. 4.7

next to the down-wind sector and *towards* the expected wind. When the wind veers you can either move across into the new down-wind sector, *CD* or, if you expect it to veer still more, you can tack in the next sector *BC*. Do not go too far out of your way looking for your wind change. The wind may not change before you reach *P*, or it may change so much that it would have freed you even had you not gone so much out of your way, or it may veer when you expect it to back and vice versa.

Very occasionally, if you are an expert meteorologist, and things go as expected you can gamble and succeed. The 1948 Santander race was won by *Eilun*. The skipper had had news of a secondary depression of which either no one else had heard, or did not know how to interpret. I do not remember the details exactly but I think that with a southerly wind we all laid for Ushant on the port tack while *Eilun* reached fast along the south coast. The depression came, swinging through SW to W and finally NW. *Eilun*, with one tack, cleared Ushant and won by hours, while the rest of us had to beat into the teeth of the gale. It was the rule of going towards where the wind is going to come from, carried successfully to its logical conclusion. But remember that for the once it works there are a dozen times when it does not.

Fig. 4.8 shows a 200 mile trip down Channel from Dover to Torbay with two changes of wind. By immediately using to the full the advantages offered by these you can reduce the distance sailed through the water from the 300 miles you would have to sail in a dead beat, to about 245 miles. Naturally every gust will not be taken for a change of wind. You must be sure that the wind has really changed; wait ten minutes or so before tacking just to make certain it is not temporary. A true shift of wind is often accompanied by a change in the force of the wind; if it freshens sharply with the change it will probably not change back again. Remember that in a depression or after a front passes the wind will always veer.

Beating in Very Light Winds   Do not, as navigator, try to insist on the helmsmen keeping on course or heading up well in light airs. This is the occasion when the course is decided by the helmsmen keeping her

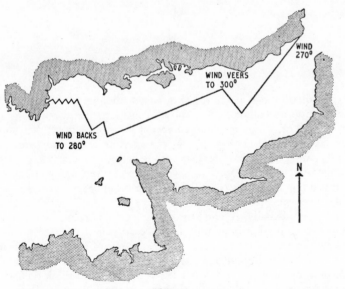

Fig. 4.8

sailing. Your navigation is not going to be thrown out very much whatever they do as the distance covered is so small, and it is better to be one mile nearer your objective after four hours' calm than to be heading up nicely on course in exactly the same place.

RUNNING

Running dead before the wind is a slow and uncomfortable point of sailing for all boats. If your course lies dead down-wind this is one of the occasions when it would pay not to go straight but to head up and gain speed on one gybe or the other. How much you can afford to alter course varies from boat to boat and depends on what sails you are carrying but the principle is as follows.

Strong to Fresh Winds Only a very small alteration of course is needed to bring the wind enough on the quarter to drive the boat forward at her best pace. Going for a distant mark the angle off course (in the region of 5°) is too small to worry you. You are almost certain to have a wind change before you reach the mark, which will bring the wind on one quarter or the other, and so allow you to go straight for your point. If not you should gybe when the mark lies about 10° to 15° to leeward.

Light Winds These conditions necessitate having the wind a little more on the quarter, but a 20° divergence from course will only cause you to sail 6 per cent further (Fig. 4.9(a)). This shows that if you are doing 4 knots you have only to do more than 4·25 knots to make the alteration of course worth while. Your increase in speed will probably be much more.

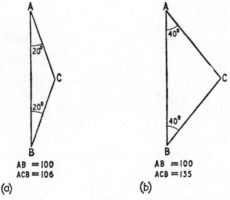

AB = 100
ACB = 106

(a)

AB = 100
ACB = 135

(b)

Fig. 4.9

Very Light Variable Winds   With these winds the important thing is to 'keep her sailing'. (Here I am referring only to racing, the cruiser will wisely be under power!) Suppose you have to head 40° off course to keep way on you will sail 30 per cent farther (Fig. 4.9(*b*)); that is to say, if you are doing 1 knot down-wind and can do more than 1·3 knots headed up, it is worth it.

To develop these arguments further one must enter into the realms of seamanship which is outside the scope of this book. Do you want to sail faster and set a mizzen-staysail? Do you want to bear away until you can carry the big genoa or a spinnaker? These are racing problems which the skipper will put to the navigator. The navigator's job is to make sure that if the distance is increased the speed is increased *more* than correspondingly. If the navigator cannot see a clear advantage in a longer route he should stick to the shorter one. Always remember, if it does drop light, those extra two miles may take an hour to cover.

TIDES

I was brought up on the East Coast where (and I think that to a lesser extent this is also true of the Solent) the winds were less important than the tides. The passage across the Thames Estuary involved a long, intricate sail through narrow channels, the Wallet Spitway, the Barrow Deep, the Black Deep, the Edinburgh Channel and the Downs, until at last, at the South Foreland, we came into open waters. Navigation was reduced to recognising the buoys, not overshooting the entrances to the channels and, because there was no choice of how to use the wind since one's direction was strictly predetermined, a desperate effort to use every half knot of help the tide could give us.

I think this prevented me for a long time from realising that, in open waters, wind is much more important than tide. If you alter course to sail close inshore at a headland you may get an extra two knots of tide for two hours, but a wind change could make you pay a very high price for this diversion. If you sail five miles farther you must *be sure* of getting at least six miles extra help from the tide. Nothing is worse than the wind dying away just as you get to your special tide (rather late) and there, more or less, you stay until it turns strongly against you.

It can be put as follows: on short trips, or when sailing in confined channels, tides are very important indeed; on long passages they should be exploited when convenient but no great weight put on them from a planning point of view. This does not in any way mean that you can ever ignore the tides in your D.R.; they must always be *plotted* with the greatest care. Thus, if the wind gives you two equally good alternatives, take the one where the tide is more favourable, but do not go out of your way tide chasing. If the wind is dead ahead and the tide at right angles to it, then it will pay to go off on the tack which brings the tide on the lee bow and not that which brings it on the weather quarter. The tide under your lee bow will help keep you nearer to your objective, and nearer to where the wind is coming from, to give you the benefit of a change. It will also increase the relative wind speed.

It is a very limited situation, however, for if the mark or point you are sailing for is far away and if there is a good sailing breeze it does not necessarily pay to take the worse tack in order to have the tide on the lee bow. In Fig. 4.10 the mark lies along the line AP due east, the wind is 10° south of east and a two-knot tide is setting from the south. Two boats sail the different tacks AB and AC for three hours and the tide carries them to B_1 and C_1 respectively. All points on the line XYZ are equidistant from B_1 and C_1. It will be seen that the line XYZ crosses the line AP at Y, 35 miles from A, so that if the mark is nearer A than Y, C_1 is nearer to the mark after three hours; if the mark is further off than Y (i.e. more than 35 miles) B_1 is nearer.

Now in this example AB is only just the better tack; if the wind veers 10° more, making AB the better tack by 20°, the wind has to be

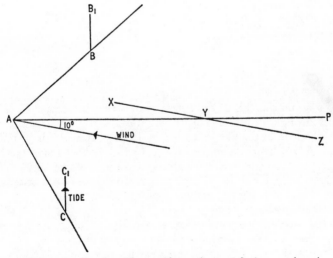

Fig. 4.10.   Speed = 6 knots; tide = 2 knots 360°; time = 3 hours' sailing

very light, the tide very strong and the mark very near before it is worth taking the worst tack just for the sake of the tide. I realise this is a theoretical argument which will normally only be brought into use when racing, but it goes to show that it is all too easy to exaggerate the importance of tides over long distances. Each case must be considered on its merits, taking into consideration the following points: the strength and direction of the tide, the time the tide will slacken or turn, the distance before you sail out of the area of strong tides, the direction of the wind, which is the better tack and by how much, the distance away of the mark and the probability of the wind backing or veering.

From a purely racing point of view the main factors affecting strategy (wind changes, tides, sailing off direct course to gain speed) are already complicated when you consider them separately: at sea you often have to take them all into account at the same time, possibly when working within a largish area of uncertainty, the possibilities then become so numerous that the navigator must keep a clear head and not get led into taking a gamble on one factor while ignoring the others.

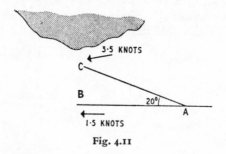

Fig. 4.11

Do not run the risk of the situation changing before your plan is carried through. In Fig. 4.11 if BC is 2 miles then go inshore and try to get the help of the tide, but if BC is 10 miles then AC is 30 miles, or probably 6 hours' sailing. In the latter case your decision to alter course has to be taken so far ahead that you must be 100 per cent certain that the wind will not drop before you get to the headland; otherwise you may find yourself kedged in a foul tide for several hours.

For short distances, however, tides are of the greatest importance. Take the simplest of all cases (Fig. 4.12); AP 5 miles, 180°; a 5-knot breeze; tide, 1 knot, direction 270°. You lay off the tide PC on your chart and steer for C (167°), and the tide will carry you to P. Naturally you must check this tide if you want to arrive at P without loss of time.

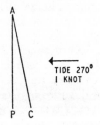

TIDE 270°
I KNOT

Fig. 4.12

John Illingworth's story (quoted on p. 45) stresses the value of back bearings for this check. If you sail away from a mark on a course of 180° the mark should remain, if there are no tides or currents, on a bearing from you of 360°. Back bearings will soon show you how much the tide is setting you off your course. As soon as you can see P clearly, you can take bearings ahead and when near you can check by a very easy method. Take any object at P, a tower, a lighthouse, chimney, etc., and make sure it is not moving against the land behind. If it is moving you are not going straight for it. If during the passage the wind drops, thus altering the speed of the boat, a new course to steer must be worked out.

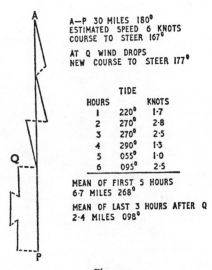

A—P 30 MILES 180°
ESTIMATED SPEED 6 KNOTS
COURSE TO STEER 167°

AT Q WIND DROPS
NEW COURSE TO STEER 177°

HOURS	TIDE	KNOTS
1	220°	1·7
2	270°	2·8
3	270°	2·5
4	290°	1·3
5	055°	1·0
6	095°	2·5

MEAN OF FIRST 5 HOURS
6·7 MILES 268°

MEAN OF LAST 3 HOURS AFTER Q
2·4 MILES 098°

Fig. 4.13

Fig. 4.13 shows a more elaborate case. *AP* is 30 miles, you start with a 6-knot breeze, estimated time 5 hours. When you reach Q the wind drops, and your speed drops to 4 knots; your estimated time for the rest of the passage becoming 3 instead of 2 hours. The D.R. is plotted for each hour and the tide correction added.

There are other considerations which may lead you to leave the direct course or the best tack, but they depend so much on circumstances that they cannot be discussed in detail. If there is a strong head wind and a big sea, it often pays to tack into a bay to get into the lee of the land and into calmer waters. On the other hand, cliffs and high land frequently have calms close inshore, and your course should be chosen to avoid these.

But let me stress once more, normally the shortest distance is the quickest way, so go straight unless you have a really valid reason for doing otherwise.

CHAPTER FIVE

REACHING
YOUR DESTINATION

LANDFALLS

Landfalls by day are seldom as easy as you think they are going to be. By night they are much easier as the characteristics of the lights identify the various points immediately. When distant land appears, misty and far away, do not listen to your crew who will start saying: 'That's the hill above Greenford, I know it well'. 'You can't mistake that: it's Steppington, the line of the cliffs is unmistakable.' 'We are too far to the west, that's Mount Olympus.' Before altering course, wait until you are sure yourself. And I mean sure; towers can look like lighthouses, lighthouses like churches, the headland which seems so conspicuous on the chart merges with the land behind, grey cliffs appear white in the sunlight—things are not what they seem.

Nor is it easy to estimate distance off. A hill that appears to be ten miles away may be five miles off on a misty day, fifteen on a clear day and twenty-five in the Mediterranean! Nor, unfortunately, are the drawings in the *Pilot* much help, at least I have never found anyone able to use them. The answer is patience and faith until you get nearer. This is all right if the coast is clear, but there are coasts where unmarked rocks and shoals lie miles out to sea, as for example the coast with the encouraging name of 'la costa de la muerte' on the NW corner of Spain, near Vigo. Here, if I was not absolutely certain, and not racing, and had to make a landfall from the west, I should heave to and wait for the lights at dusk to give me a fix.

For some strange psychological reason, when land is in sight, even if not recognisable, navigators rarely think of using celestial navigation. They are capable of continuing in uncertainty for some hours when often they could easily resolve their doubts in a few minutes by taking a Sun sight.

FOG OR BAD VISIBILITY

Having kept your usual accurate D.R. during a forty-mile night passage you find yourself in fog with a headland to round. In Fig. 5.1,

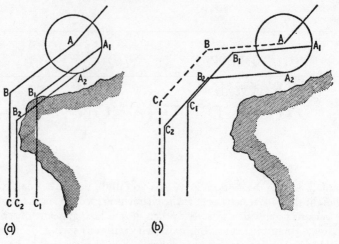

Fig. 5.1

A is your D.R. plot. Give yourself a good large area of uncertainty, say 8 per cent instead of 5 per cent of the distance sailed, and then take the two worst points on this circle. A_1 is the most easterly point and A_2 the most southerly. Now it will be seen in (*a*) that your normal fair-weather course, *ABC*, while all right from *A*, will run you straight ashore if sailed from A_1 or A_2. Fig. 5.1(*b*) shows a course (265°, 8 miles; 223°, 7·5 miles; 180° onwards) which, whether sailed from A_1 or A_2, takes you clear of the headland. Work out this course and then sail it, plotting it from *A*. There are three points to remember:

1. This course must be corrected for tides or currents before being given to the helmsman, so that if the tide is running NE and E round the headland, the courses to steer and the distances to sail through the water might be 270°, 10 miles; 235°, 9 miles and then 190°.

2. Your area of uncertainty is growing continually, and if in an area of strong tides, rapidly; so if the fog does not lift you should keep out in open water after rounding the headland.

3. All this presumes no outside help; you will of course have been plotting radio-bearings, listening to fog-signals and using your echo-sounder continually, but do not cut inside your 'safe' course unless you are certain that it is safe to do so.

Fig. 5.2 gives another example of using your area of uncertainty to avoid difficulties. Your estimated position *A* is three miles from the headland when a passing steamer enables you to put the visibility at about three miles. You cannot, therefore, be in the shaded area of sea

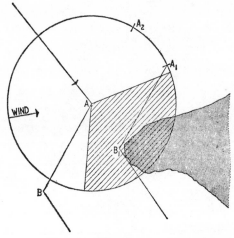

Fig. 5.2

or you would see land. The wind is W by S, *AB* is the highest you can lay on the starboard tack. If you are at A_1, and go any further on your present course, you will have to beat to round the point. So to avoid this you head up to course *AB* until you are certain that the headland bears east, before bearing away. Here a lead, or echo-sounder, should tell you roughly which half of the circle you are in, unless the bottom is very flat or very uneven. If the weather thickens, you must also make sure you will clear the headland if you are at A_2.

These are examples of the way in which your area of uncertainty can be used to help you in fog, bad visibility or when you are 'lost'. Every situation will need its own solution, but these are the occasions when accuracy and experience tell, and ten minutes' work on the chart can save hours of sailing. The one essential is to make sure that your area of uncertainty is large enough. This has been discussed on p. 44 but it is a subject which I approach with the greatest diffidence and cannot lay down any rules.

By this method of areas of uncertainty you should never feel able to say 'I am completely lost'. Firstly, this is not true, you must know if you are in the North or the Irish Sea! If in the latter you must be able to work out whether you are 100 miles or 50 from any given point. Some area must be calculable, however large, from which you can start to work, and which can be reduced by radio bearings, etc. And here I must urge the navigator to take to celestial navigation. A glimpse of the Sun through clouds will permit a sight and a position line which will, at the least, limit your area considerably. Another

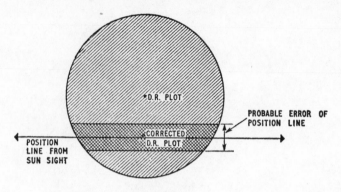

Fig. 5.3

glimpse, three hours later, will be enough to give you a fix of some sort. Of all methods of navigation it is the least used by the amateur although, with very little practice, it is one of the most useful. Fig. 5.3 shows how a single position-line from the Sun can reduce your area of uncertainty. This, after sailing 200 miles, has a radius of 10 miles and an area of 314 square miles. The position line reduces it to the area cross-hatched which is about 80 square miles.

PILOTAGE

Pilotage is the name given to navigation in narrow waters and harbours where a large ship carries a pilot. Most aids to navigation are useless; Consol, radio bearings, celestial navigation and log can all be put aside. You are left with compass, chart, echo-sounder (or lead) and eye. Some fortunate people have a gift for judging distances by eye (I have not) and, by looking at the chart and then at the harbour in front of them, can locate themselves at once. Others must use the hand bearing-compass until very close in. But everyone should study the chart before arriving. Get clear in your mind whether the harbour moles run north and south, or east and west; make a mental note that your compass course for entry will be, say, 180° and that after that you will turn on to 245°. Read the pilot book before entering, if you do not know the harbour, and see if there are leading lines. These are particularly useful, and are found in a great many French ports, leading you up long narrow harbours such as Morlaix or L'Aberwrach. They give you the greatest sense of security because while they are in line you *know* you must be all right (Fig. 5.4).

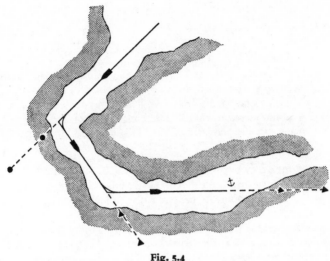

Fig. 5.4

At night, large harbours are most confusing. The navigational lights are mixed up with masses of lights on shore, advertisements, headlights of cars, lights on ships, quays, etc., so that the more you have directions and distances in your head the better. Remember also that many harbours, particularly on the Continent, are in the course of being enlarged, and thus the lights on the ends of inside moles, etc., may be changed at short notice. Keep a sharp lookout and have a torch handy on deck for glancing at the chart.

While entering harbour is more difficult at night, navigation in narrow channels, such as the Thames Estuary, is more difficult by day. The light-buoys can be picked up earlier at night, and their character-istics make them easily distinguishable one from the other. By day, if you are not 100 per cent sure which buoy is which *go and read what it says*. Sometimes this applies even to light-vessels. One year, coming back from the Hook of Holland, we were just going to round the *South Goodwin* light-vessel when we saw *East Goodwin* written on the side of it and had to alter course 90° in a hurry. I have heard that when navigating in the shallow winding channels of the river Plate, it is better to read the buoys and tick them off as one goes along, and I am sure it is true of such rivers as the Scheldt, Weser and Elbe. It is better to be safe than sorry.

Study carefully beforehand the system of lights in any channel much used by big ships since it may be fairly complicated. If you

73

consider, for example, the Needles Channel you will see that the Needles light has:

1. A red sector to the SE, covering a $5\frac{1}{2}$ fathom shoal, to the east of which the light is obscured.
2. A red sector to the NW covering the Shingles bank. This red sector must be sailed through in the channel itself.
3. A white sector from SE to W leading up to the channel.
4. A white sector to the ENE leading up to Hurst Point.
5. A green sector covering Warden Ledge and How Bank along the island shore.

Hurst Point has two towers with lights which, when in line, lead into the channel. Once through the narrows at Hurst the higher light has a narrow red sector down the centre of the Solent covering the Solent Banks.

When you realise that, in addition to all this, there are a number of lit buoys, it will be clear that to sail unprepared into such a galaxy of lights would be extremely muddling, while ten minutes at the chart beforehand, with a few notes, would enable you to make full use of all the assistance provided.

For all pilotage you should have the chart on deck for constant reference. A plastic-covered board into which you can slip the folded chart will keep it dry, and a chinagraph pencil will enable you to draw in bearings. The objects on which you are taking bearings are so close that a slight inaccuracy in drawing them will not be very important. Remember to study also the tides before entering harbour; you should know the rise and fall, and what the state of the tide is at that particular moment. Thus if your lead or echo-sounder tells you that you have three fathoms under you, you will know at once if you can anchor safely, or will wake up next morning to find yourself high and dry.

CHAPTER SIX

GOING FOREIGN

Venturing into new waters poses few problems for the navigator provided he reads and studies the necessary literature. The required volumes are:

1. The *Pilot* for the areas to be visited.
2. The *Admiralty List of Lights* for the same areas.
3. The *Admiralty List of Radio Signals*, Vols. II, III, V.

You must of course take all the necessary charts. The Admiralty Hydrographic Department does not always publish very large-scale charts of the coasts of countries which have their own hydrographic departments. If, therefore, English charts are not available in great enough detail, get those published by the country in question. Remember that soundings on foreign charts are usually in metres (except American charts on which fathoms are used); always check up on this if you have a mixed lot of charts on board; while you will float in two fathoms you may not in two metres!

It must be remembered also that the information may not be as accurate, either on the charts or in the *Pilot* or the list of lights, on some less frequented coasts, as that for home waters. Recently, for instance, new harbours have been built in Spain, and for some time were not marked on the charts; the blaze of lights at night was extremely confusing. On the other hand, I have heard that the lights on the west coast of Mexico are liable either not to be lit at all, or to be quite different from those listed. Therefore, a certain amount of caution is sometimes a very necessary thing.

The first chapter of each *Pilot* should be studied with great care, noting the tides, currents, frequencies of gales, prevailing winds, local winds, etc., before visiting the coast. The details of the coasts and ports should be read day by day. Do not imagine that the *Pilot* makes boring reading! Here is a description of a gale in the Orkneys:

In the terrific gales which usually occur four or five times in every year all distinction between air and water is lost, the nearest objects are obscured by spray, and everything seems enveloped in a thick smoke; upon the open coast the sea rises at once, and striking upon the rocky shores rises in foam for several hundred feet and spreads over the whole country.

The sea, however, is not so heavy in the violent gales of short continuance as when an ordinary gale has been blowing for many days; the whole force of the Atlantic is then beating against the shores of the Orkneys, rocks of many tons in weight are lifted from their beds, and the roar of the surge may be heard for twenty miles; the breakers rise to the height of 60ft, and the broken sea on the North Shoal, which lies 12 miles northwestward of Costa Head, is visible at Skail and Birsay.

OCEAN CROSSINGS

Crossing oceans raises different problems for the navigator from those of every day coastal sailing. The most important difference is that while up to now he has been able to get along without celestial navigation this is no longer possible. A good working knowledge of celestial navigation is essential. I do know of people who have started, knowing nothing, and arrived having learned all they need to know, but it is preferable to have had some practice before embarking on a trans-oceanic passage.

Even more than in coastal sailing the planning done beforehand is of the greatest importance. The navigator intending to sail must read and study enough to decide when, and which way, to go; and enough not to be caught out by foreseeable difficulties or dangers.

The following publications must be obtained. *Ocean Passages for the World*, the pilot charts for the ocean in question and the *Pilot* for the coasts where landfall is likely to be made; these are published by the Hydrographic Department of the Admiralty. There is one pilot chart for each month of the year for the various areas; they give the percentage of winds from different directions, the average number of winds of gale force during the month and the ocean currents. A careful study of these will ensure your taking the right route at the best time. Here can be found the trade winds, the doldrums, the westerlies and other prevailing winds. Here is the information about the hurricane season in the West Indies, on typhoons in the Pacific and monsoons in the Indian Ocean. Here are the great currents, the Gulf Stream, the Labrador current, the Equatorial current and others.

Before, however, deciding on routes a slight knowledge of great-circle sailing is necessary. A loxodrome is a line on the surface of a sphere which makes equal oblique angles with all meridians. On Mercator's projection, the projection used for ordinary charts, a loxodrome is as a straight line, usually called the rhumb-line, or in this book, direct course (Fig. 6.1). Now it would be perfectly possible to sail a rhumb-line course from New York to Lisbon (or between any other two points on the Earth's surface). As they are almost on the same parallel of latitude, if you sail east from New York you will arrive at Lisbon. But this is not the shortest way. You can prove this satisfactorily to yourself with a globe and a piece of string. Hold the string on the two cities, pull it as tight as possible and you will see that it does not follow the parallel of latitude on which they both so nearly lie.

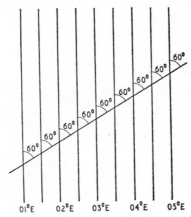

Fig. 6.1.   Loxodrome on Mercator's projection

The shortest distance between two points on the Earth's surface is always part of a great circle. A great circle is any circle having as its centre the centre of the Earth and its radius the distance from the centre to the surface of the Earth. The equator and the meridians are great circles, but other parallels of latitude are not, because the centre of a circle formed by a parallel of latitude lies either north or south of the centre of the Earth. The angles formed by the meridians and any great circle (except the Equator) are not equal; therefore a great circle will appear on Mercator's projection as a curve.

How then, if you wish to know the shortest way from New York to Lisbon, can you find it out? The easiest method is to buy a gnomonic projection chart from an Admiralty chart agent. On this projection any straight line is part of a great circle. Rule a line from the point of departure to the point of arrival and take off the geographical coordinates of a series of positions along this line. Every five degrees of longitude should suffice. Mark these positions on your ordinary Mercator chart. The effect is shown in Fig. 6.2. Courses and distances can now be taken off in the normal way.

Now the great-circle course is the shortest but it may not be the best

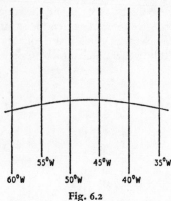

Fig. 6.2

or the quickest and must be studied in conjunction with winds and currents. Crossing from New York to Lisbon in winter it might be convenient to follow the great-circle course, but in summer it would certainly pay to go further north in search of westerly winds rather than risk being becalmed north of the Azores. On the other hand the passage from Boston to Norway will probably be made to the south of the great circle to avoid risk of icebergs.

The crossing from east to west presents other problems. You can choose to follow the great circle, which more or less involves beating 3,000 miles against the Gulf Stream, or you can go the long way round, 5,000 miles odd, down to the Canaries and then across in the trade winds to the West Indies; much longer to be sure, but with the certainty of favourable winds and currents.

In addition to studying the crossing, all possible information must of course be obtained about the coasts where you expect to arrive (make a good wide selection too; I once set sail from Cannes to go to the West Indies and arrived in Tunis!). You must take the relative *Admiralty List of Lights, Pilots*, charts, *Admiralty List of Radio Signals*, Consol charts if available, and, if possible, for reading, any accounts of similar trips by other yachtsmen. If I were going anywhere really off the beaten track, like Greenland or Patagonia, I would certainly try my level best to meet and talk over the problems with someone who had already sailed there.

The only addition to your normal equipment is a sextant, a deck watch and a wireless capable of receiving short wave if necessary. Remember you can only avoid buying a chronometer by having a good reliable wireless. You will also need tables and an almanac. Such modern electronic equipment as Loran and Omega receivers is not within the scope of this book. World-wide accurate positioning at the touch of a switch is now available, but batteries can fail, or equipment break down, and no serious navigator will risk long voyages without the equipment and ability to take astronomical sights.

While the work before sailing is more important than for short passages, the trip itself is a holiday from the navigator's point of view unless racing. In the first week or so D.R. is of little use, the position can be plotted from sights once or twice a day and the weather forecasts listened to, but that is about all. For plotting it is useful to carry plotting charts, since it is difficult to work accurately on the very small-scale charts of oceans. These are published by the Admiralty Hydrographic Department and are sheets of paper marked only with meridians and a scale by which you can draw your own appropriate parallel of latitude. The position for the day can then be transferred to the chart.

Only as you approach land does the work become increasingly important. Six or seven hundred miles from land you must start plotting your D.R. with more care, because it may be overcast for three or four days; your area of uncertainty may become enormous but you must have some idea of where you are. But, even if you can get sights, about three hundred miles off land you should revert to ordinary coastal practice: D.R. plot every hour, checked by sights, confirmed by Consol if necessary, and, eventually, as they come into range, reconfirmed by bearings from radio beacons; until that glorious moment comes, which can never fail to thrill the amateur navigator, when you say: 'At dusk we should see the light', and at dusk, there on the bow is the double flash of the Bishop Rock. It really works!

APPENDIX

COMPASS DEVIATION, SWINGING AND ADJUSTMENT

by MARIO BINI

BASIC ASSUMPTIONS

The basic assumption in the study of magnetism aboard is that the iron affecting the compass is of only two types: *soft iron*, which has no magnetism of its own, but instantly acquires it by induction from the Earth's field; and *hard iron* (e.g., steel) which possesses a permanent magnetism of its own, thus producing a magnetic field completely independent of the Earth's field.

This assumption is not entirely true, as there are 'intermediate' kinds of iron which do not acquire induced magnetism instantly but with a certain delay. There is no way of taking this iron into account, at least theoretically, but p. 92 explains how it is dealt with if it is present on board.

Both soft and hard iron have a disturbing effect on the compass, an effect which varies with the ship's heading and which causes the

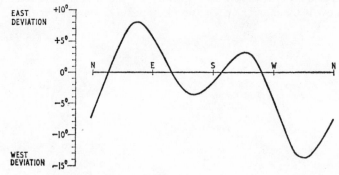

Fig. A.1

compass to be deflected by a certain amount east or west of its correct direction, i.e. the magnetic north. The amount the compass is deflected is called *deviation* and is measured by swinging the compass as described on p. 93.

Since deviation depends on the ship's heading, when represented graphically it is always plotted against compass headings, as shown in Fig. A.1.

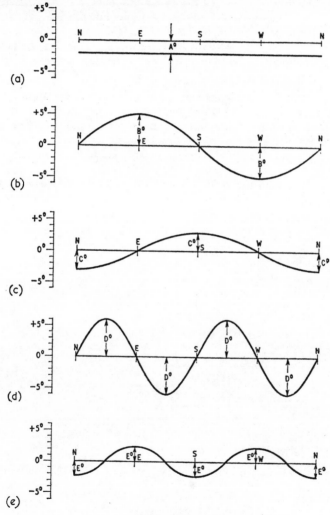

Fig. A.2

Appendix

Theory demonstrates, and practice fully confirms, that the deviation curve, no matter how complex it might appear, is the sum of five component 'simple' curves. These curves are:

1. A constant deviation, represented by a horizontal line (Fig. A.2(*a*)).
2. A sinusoid, of which the maximum value is $B°$ (Fig. A.2(*b*)).
3. A co-sinusoid (that is to say, a sinusoid shifted by 90°) of which the maximum value is $C°$ (Fig. A.2(*c*)).
4. A double sinusoid with $D°$ as maximum value (Fig. A.2(*d*)).
5. A double co-sinusoid with $E°$ as maximum value (Fig. A.2(*e*)).

These component curves are produced by the action of five forces into which the ship's magnetic field can be broken down. In order to determine the values of $A°$, $B°$, $C°$, $D°$, $E°$ the compass must be swung, the deviation curve drawn, and the deviations read off for the eight cardinal and intercardinal headings (0°, 45°, 90°, 135°, 180°, 225°, 270°, 315° 330°); the deviation at each of these headings is indicated by $d_{0°}$, $d_{45°}$, $d_{135°}$, etc.

From these deviations, $A°$, $B°$, etc., can easily be calculated using the following formulae, in which easterly deviations must be given a positive sign $(+)$ and westerly deviations a negative sign $(-)$:

$$A° = \frac{d_{0°} + d_{45°} + d_{90°} + d_{135°} + d_{180°} + d_{225°} + d_{270°} + d_{315°}}{8}$$

$$B° = \frac{d_{90°} - d_{270°}}{2}$$

$$C° = \frac{d_{0°} - d_{180°}}{2}$$

$$D° = \frac{(d_{45°} + d_{225°}) - (d_{135°} + d_{315°})}{4}$$

$$E° = \frac{(d_{0°} + d_{180°}) - (d_{90°} + d_{270°})}{4}$$

AN EXAMPLE OF HOW TO DETERMINE A°, B°, C°, D°, E°

Applying these formulae to the deviation curve in Fig. A.1, in which:

$$d_{0°} = 7°·5 \text{ W} \qquad d_{180°} = 1°·5 \text{ W}$$
$$d_{45°} = 5°·5 \text{ E} \qquad d_{225°} = 3°·0 \text{ E}$$

$$d_{90°} = 5°\cdot5 \text{ E} \qquad d_{270°} = 4°\cdot5 \text{ W}$$
$$d_{135°} = 2°\cdot5 \text{ W} \qquad d_{315°} = 13°\cdot5 \text{ W}$$

and remembering that easterly deviations must be given a plus sign and westerly a minus sign, we obtain:

$$A° = \frac{-7\cdot5 + 5\cdot5 + 5\cdot5 - 2\cdot5 - 1\cdot5 + 3\cdot0 - 4\cdot5 - 13\cdot5}{8} = -2°$$

$$B° = \frac{+5\cdot5 + 4\cdot5}{2} = +5°$$

$$C° = \frac{-7\cdot5 + 1\cdot5}{2} = -3°$$

$$D° = \frac{(+5\cdot5 + 3\cdot0) - (-2\cdot5 - 13\cdot5)}{4} = +6°$$

$$E° = \frac{(-7\cdot5 - 1\cdot5) - (+5\cdot5 - 4\cdot5)}{4} = -2\cdot5°$$

These are the component deviations of the curve represented in Fig. A.1, and are shown in the figures from A.2(*a*) to A.2(*e*); if in fact the five component curves are drawn on the same graph and their sum is graphically drawn along vertical lines, the original deviation curve of Fig. A.1 would again be obtained.

COMPONENT DEVIATIONS ON NORMAL SHIPS

The study of magnetism demonstrates that deviations $A°$ and $E°$ are due to certain types of asymmetrical soft iron which are not found on 'normal' ships, meaning by 'normal' those ships which are symmetrical (some, such as aircraft-carriers, are not) and which have the compass placed amidships, on the line of the keel.

From this point of view wooden yachts are certainly 'normal'; the only soft iron on board is in the engine (and the keel, if of iron), which anyway is generally placed well below the compass. This greatly reduces the disturbance in the horizontal plane, which is the most important: the vertical component has no deviating effect on the compass (it would only tilt the compass rose but not deviate it) as long as the boat is horizontal, but can give rise to a deviation when the boat is heeled (see p. 92).

Experience shows that, both on board ships and yachts, deviations $A°$ and $E°$ are negligible, and, for this reason, no device is provided to correct them. Should they be found to have a significant value the only thing to do is, either find another position for the compass, or remove the offending iron.

From now on, therefore, we shall take it that the deviation curve is due to three component deviations only:

1. The sinusoidal B.
2. The co-sinusoidal C.
3. The double sinusoidal D.

THE EFFECT OF MAGNETS AND SOFT IRON ON THE COMPASS

Let us now see how a compass behaves if soft iron and hard iron (e.g., a magnet) are placed near it. Let us start with a magnet placed in the

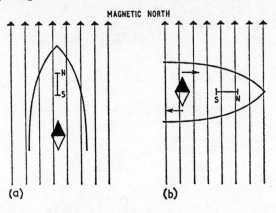

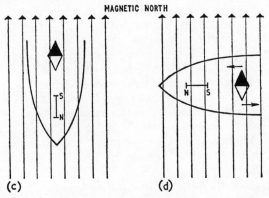

Fig. A.3

fore-and-aft line, with its south polarity towards the compass. When the ship is heading north (all headings referred to are magnetic headings) the deviating effect on the compass is nil (Fig. A.3(*a*)), because the magnet will only increase the Earth's magnetic field without altering its direction.

If the ship is heading east (Fig. A.3(*b*)) the north end of the compass will be attracted by the south polarity of the magnet and the south end of the compass repulsed: the compass will turn clockwise and the deviation will be easterly. If the boat is heading south (Fig. A.3(*c*)) the magnet's deviating effect is again nil, as it simply reduces the Earth's magnetic field without altering its direction. If the boat heads west (Fig. A.3(*d*)) the south polarity of the magnet will attract the north end of the compass and repulse the south end: the compass will turn anti-clockwise and the deviation will be westerly.

A magnet placed as described, therefore, produces easterly deviations on headings from north to south passing through east, and westerly deviations on headings from south to north passing through west. The maximum deviations are reached when the heading is east or west as the action exerted by the magnet is then at 90° from the compass needle.

If the magnet were reversed, a contrary effect would be obtained: westerly deviations on headings in the eastern semi-circle and easterly in the western.

It will be obvious that by varying the strength of the magnet and/or its distance from the compass, the amount of the deviation produced can be varied accordingly. It can now be seen that, by placing a suitable magnet in the fore-and-aft plane, a deviation such as

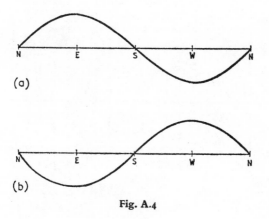

Fig. A.4

that shown in Fig. A.4(*a*) can be produced, or, reversing the magnet, such as that in Fig. A.4(*b*). Now these are deviations of exactly the same type as deviation *B* (Fig. A.2(*b*)); *if, therefore, we put, in the fore-and-aft plane, a magnet of appropriate force at the right distance from the compass, we can produce a deviation which, being equal but contrary to B, annuls it.*

If we put a magnet athwartships, alongside the compass, we obtain a result very similar to the one we have already looked at, but rotated by 90°: the magnet has no deviating effect when the boat heads east and west (Fig. A.5(*a*)), its only action being to increase or reduce the Earth's magnetic field without altering its direction. The maximum deviations are reached when the compass needle is at 90° to the magnet,

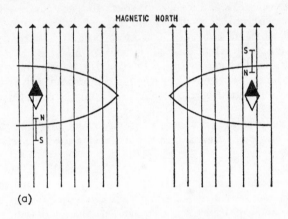

(a)

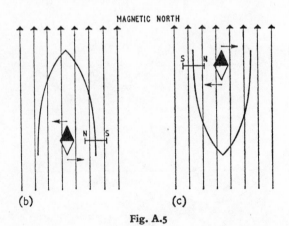

(b)　　　　(c)

Fig. A.5

on headings north (Fig. A.5(*b*)) and south (Fig. A.5(*c*)), the deviation being west when heading north and east heading south.

We can thus obtain a deviation which is nil for headings east and west and reaches its maximum value on headings north and south.

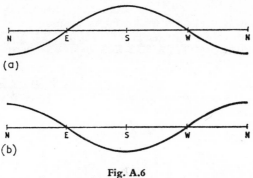

(a)

(b)

Fig. A.6

This deviation is shown in Fig. A.6(*a*), or, with the magnet reversed, in Fig. A.6(*b*). It will be seen that it is of exactly the same type as deviation $C°$ (Fig. A.2(*c*)); *if, therefore, we put a magnet of adequate strength at the right distance athwartships, we can produce a deviation which, being equal but contrary to C, annuls it.*

(The rest of this section does not apply to wooden yachts, but is included for general interest and for those who may have to deal with compasses aboard steel yachts or ships.)

Let us see what happens if we put some soft iron, such as two spheres, abeam of the compass. As said on p. 80, soft iron has only induced magnetism: this means that, with the action of the Earth's magnetic field each sphere will become magnetised (Fig. A.7); and the half

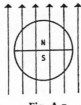

Fig. A.7

which looks north will be charged with north magnetism and the other half with south magnetism. When the ship is on the magnetic headings north, south, east and west (Fig. A.8) no deviation will be produced by the two spheres, because on headings north and south both ends of the compass are repulsed by equal forces, while on headings east and west the only effect of the spheres is to increase the Earth's magnetic field without altering its direction.

On the intercardinal headings (NE, SE, SW, NW), however, a deviation is produced, as can be seen in Fig. A.9 which represents a boat heading NE. Consider the starboard sphere first: its upper half (having north magnetism) is nearer than its lower half (having south magnetism) to the south end of the compasss, with the result

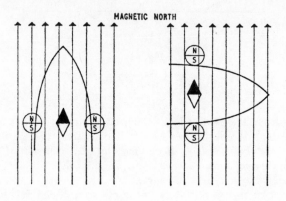

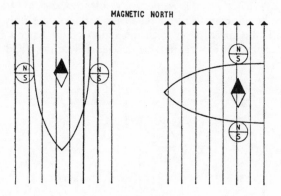

Fig. A.8

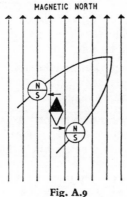

MAGNETIC NORTH

Fig. A.9

that the south end of the compass is attracted by the starboard sphere. Similarly with the port sphere, the lower half of which will have a stronger effect than its upper half (being nearer), with the result that the north end of the compass will be attracted by it. Since the forces of both spheres tend to turn the compass anti-clockwise a westerly deviation is produced. Of course, varying the size of the spheres and/or their distance from the compass, the amount of the deviation produced can be varied.

A similar thing happens on the other intercardinal headings, and the conclusion is that, by putting the two spheres (cylinders can also be used) of soft iron on either side of the compass, a deviation is produced which is zero on cardinal headings and reaches maximum values on intercardinal headings. This deviation always starts with a westerly deviation as shown in Fig. A.10. Fortunately, due to the

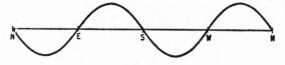

Fig. A.10

fact that ships are longer than they are wide, deviation $D°$ always starts with an easterly deviation in the NE quadrant and, therefore, *spheres of appropriate size, at the right distance, produce a deviation which, being equal but contrary to $D°$, annuls it.*

89

HOW COMPONENT DEVIATIONS CAN BE ANNULLED

We have seen that the three component deviations likely to be found aboard a normal ship can be annulled by placing near the compass:

1. Longitudinal magnets to annul deviation $B°$.
2. Transverse magnets to annul deviation $C°$.
3. Soft-iron correctors to annul deviation $D°$.

Indeed, compass binnacles in all types of ships are only equipped with the devices necessary for correcting deviations $B°$, $C°$ and $D°$ (with the possible addition of a cradle for correcting heeling error (see p. 92)).

Fortunately, another big simplification applies to wooden yachts: deviation $D°$ does not exist for them, as it is due to types of iron (mainly steel decks) which wooden yachts do not have. This is the reason why yacht compasses never have soft-iron spheres, but only longitudinal and transverse containers to hold the magnets. These containers are usually little tubes welded to the base of the compass mounting.

Thus of the five original component curves, only two, as in the tale of the ten little nigger boys, are left in a wooden yacht, since $A°$, $D°$ and $E°$ should be negligible. (If deviation $A°$ is found to occur, make sure that this is not due to a misalignment of the lubber line of the compass, which would produce a constant deviation just like $A°$.)

DEVIATIONS ON WOODEN YACHTS AND THEIR COMPENSATION

As far as wooden yachts are concerned we now know that the deviation curve, no matter what it is like, can be considered as the sum of deviations $B°$ and $C°$, as is shown in Fig. A.11. The important thing to notice in Fig. A.11 is that the deviation appearing on the cardinal headings east and west is due to $B°$ only, while on headings north and

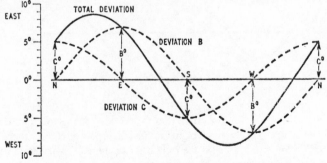

Fig. A.11

south to $C°$ only: *this possibility of isolating $B°$ and $C°$ is the key to compensation.*

Once the boat has been put on heading north, deviation $C°$ is annulled simply by placing transverse magnets until the compass reads $0°$; deviation $B°$ is in turn annulled by putting the boat on heading east and placing longitudinal magnets until the compass reads $90°$. These two operations can be started indifferently from any cardinal heading and therefore B can be dealt with before C or vice versa.

Having annulled the deviation on two adjacent headings, it is a good rule to go on turning and see what happens on the other two cardinal headings, in this case south and west. If on these headings the compass reads exactly $180°$ and $270°$ nothing, of course, has to be done. If, however, a deviation appears (this will happen if deviations $A°$ and $E°$ exist; although generally small these two deviations, added together, could reach a noticeable value) reduce or increase the magnets until the deviation is *halved*. The magnets to be used during these operations are those at right-angles to the compass needles: this can be remembered by keeping in mind the fact that magnets parallel to the needles would only increase or reduce the Earth's magnetic field without producing any deviating effect (see for instance Fig. A.3(a)).

The compass is now adjusted and the only thing to do is to swing it in order to ascertain that the deviations have, actually, been eliminated. If the adjustment has been successful, the residual deviation on any heading should not exceed $1°$. This will allow the compass to be considered free from deviations, as an error of less than one degree can be completely disregarded on a yacht.

If, however, the residual deviations have a value which cannot be ignored, two alternatives are possible: either repeat the adjustment, or leave things as they are, drawing up a compass deviation, or correction, card for the remaining deviations.

INTERMEDIATE IRON

As said on p. 80, there may be 'intermediate' iron on board, which does not get its magnetism from the Earth's field instantly, but acquires it slowly and also loses it slowly.

If a ship alters course after being on the same heading for a long time, this intermediate iron will, for a certain time, remain in the same magnetic conditions as on the former heading, thus altering the whole situation. This 'certain time' can be anything from minutes to hours, depending on the degree of 'intermediateness' of the iron concerned.

This is the reason why, theoretically, there is no way of taking this iron into account; from the practical point of view, however, there are two procedures that can be carried out in order to minimise possible trouble from this source.

1. Before adjusting the compass, the boat should be left free to swing at anchor, so that this semi-permanent magnetism cannot build up; if this is impossible, and the boat has been moored for a long time on the same heading, she should be kept moored for some hours on the opposite heading.
2. Check the compass accurately (with the hand bearing-compass or Sun's azimuth) when either altering course after a long passage always on one heading, or putting to sea if the boat has been moored for a long time on the same heading.

HEELING ERROR

We have so far dealt with horizontal forces, but vertical forces can also affect the compass. Vertical forces have no deviating effect on the compass when the ship is upright but, when the ship heels, the vertical force Z (Fig. A.12) can be broken down into a vertical component Z_1

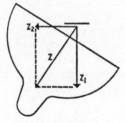

Fig. A.12

which has no deviating effect (it would only tend to tilt the rose) and a horizontal component Z_2 which, working in the horizontal plane, causes a deviation.

This deviation cannot be ignored, especially in sailing yachts. Furthermore, this deviating force works in one sense when the boat heels on one side and in the opposite sense when the boat heels on the other side. If the rolling period of the boat were equal to that of the compass, the rose would swing rhythmically, thus making steering quite impossible.

To correct this deviation, vertical magnets have to be placed under the compass to annul the vertical field: to do this the compass must be removed and a special device, called a 'heeling error instrument', used. This operation, therefore, will have to be done by a professional adjuster. Normally no compensation is necessary, but it is important to have it checked; it can be done on any heading and once done is done for ever unless some major change, such as the installation of a new engine, is carried out.

HOW TO PUT A BOAT ON A GIVEN MAGNETIC HEADING

First mark the position chosen for adjusting the compass on a large scale chart, then read from the chart the magnetic bearings of at least two landmarks referred to the position chosen. These landmarks should be approximately at 180° from each other and as distant as possible to avoid parallax errors. It is convenient to have two landmarks available because in certain headings one may not be visible, for instance when near the bow and therefore hidden by the mast.

To put the boat on a desired magnetic heading recourse is made to relative bearings: if, for example, the magnetic bearing of the landmark is 128° and the boat has to be put on a heading of 180°, she must be warped until the landmark bears 308° relative, i.e. 52° on the port bow (Fig. A.13). Compass swinging is greatly simplified by preparing beforehand a list of the magnetic headings and the corresponding relative bearings of the two landmarks chosen (Table A.1).

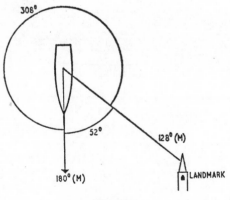

Fig. A.13

Table A.1

Magnetic heading	Relative bearings	
	Landmark A (mag. brg. 128°)	Landmark B (mag. brg. 306°)
0°	128°	306°
30°	98°	276°
60°	68°	246°
90°	38°	216°
120°	8°	186°
150°	338°	156°
180°	308°	126°
210°	278°	96°
240°	248°	66°
270°	218°	36°
300°	188°	6°
330°	158°	336°
360°	128°	306°

The best instrument for measuring relative bearings is a pelorus (Fig. A.14), if possible on a tripod so that it can be placed in the cock-

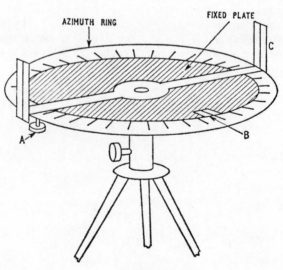

Fig. A.14

pit, exactly amidships, with its lubber line (B) in the fore-and-aft plane. The line of sight through the alidade (C) must fall on the middle of the mast when the alidade is on B, and on the centre of the stern when rotated by 180°. The pelorus is the ideal instrument for compass swinging and adjusting, especially if the alidade can be fixed by means of a screw (A) on the azimuth ring, thus allowing the alidade to be locked to the magnetic bearing of the landmark. Once this is done, the azimuth ring is rotated until the landmark is viewed through the alidade, when the boat's magnetic heading can be directly read at B.

Unfortunately, a pelorus is not a very common instrument, but a sextant can be used in its place. Mark the fore-and aft plane of the ship by placing one mark (I advise coloured Scotch tape) in the middle of the mast and another on the centre of the after edge of the coach roof (Fig. A.15): thus from the cockpit the observer has a precise line

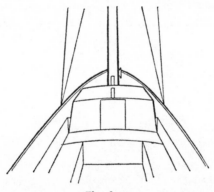

Fig. A.15

of reference. Holding the sextant horizontally the relative bearings can now be measured with great accuracy, by bringing the image of the landmark in correspondence with the two marks or vice-versa. As the sextant cannot measure angles of more than 110–120°, it is convenient to transform the relative bearings, indicated in Table A.1, into port and starboard relative bearings, as shown in Table A.2. A third landmark, approximately 90° from A and B, may prove most useful when either A or B are ahead and not visible. If necessary this 'auxiliary' landmark can be chosen on the spot and its magnetic bearing found by measuring with the sextant the horizontal angle between it and A or B and then adding (or subtracting) this angle from the magnetic bearing of A or B.

Table A.2

Magnetic heading	Relative bearings	
	Landmark A (*mag. brg.* 128°)	Landmark B (*mag. brg.* 306°)
0°	—	54° port
30°	98° strbd.	84° port
60°	68° strbd.	—
90°	38° strbd.	—
120°	8° strbd.	—
150°	22° port	—
180°	52° port	—
210°	82° port	96° strbd.
240°	112° port	66° strbd.
270°	—	36° strbd.
300°	—	6° strbd.
330°	—	24° strbd.
360°	—	54° strbd.

PRELIMINARY OPERATIONS

Before starting compass swinging, or adjusting:

1. Choose a well-sheltered place, free from any possible magnetic disturbances such as electric lines, pipes, cranes, large ships, etc.
2. Read from the chart the magnetic bearings of two landmarks referred to the position chosen.
3. Prepare a list as in Table A.1 if using a pelorus, or as in Table A.2 if using a sextant.
4. Remove all existing magnets and make sure that no iron, such as tools, etc., are near the compass; if you have magnets for adjusting the compass, put them on the bow.
5. Take enough people aboard to: warp the boat, read the compass and measure the relative bearings.
6. Moor the boat at three, or better, four, points in order to warp her easily and to be able to keep her steady.
7. Last, but not least, never try to swing, or adjust, the compass if there is wind, swell or current; you will not be able to keep her steady and you will only lose time and probably temper as well.

COMPASS SWINGING

Compass swinging, as we have seen, is done for two different reasons:

1. To measure the deviations with the object of determining the values of $A°$, $B°$, $C°$, $D°$ and $E°$ in order to make sure that the compass has been placed in a good position; this operation can be done and then never repeated unless the boat has undergone important structural changes, or if the compass has been moved. In these cases heeling error must also be checked.
2. To measure the residual deviations after adjustment.

In both cases it is enough to measure the deviations every 30°; the values so obtained are plotted on graph paper and a smooth curve drawn connecting up the various points.

From this curve you extract:

1. In case (1) the values of deviations for cardinal and inter-cardinal headings from which $A°$, $B°$, $C°$, $D°$ and $E°$ are calculated (see pp. 82–83).
2. In case (2) the values of residual deviations to be taken into account when drawing up the compass deviation card.

As the deviation curve is drawn on ruled paper it is not necessary for the values of deviation to be measured exactly on headings 0°, 30°, 60°, etc., but just near enough to have evenly spaced values on the graph; to do this the boat is warped approximately on to these headings. When she is steady the relative bearings are measured and a 'stop' given to the man reading the compass: from the angle measured, the magnetic heading is found, and the difference between the magnetic heading and the compass heading gives the deviation for that heading.

If I may give some personal advice, never use formulae, but try to remember that if the compass reads less than the magnetic heading the deviation is east and vice versa. If you already know the rhyme:

> Longitude west Greenwich time best,
> Longitude east Greenwich time least,

adapt it as follows:

> Deviation west compass reads best,
> Deviation east compass reads least.

COMPASS ADJUSTING

In contrast to compass swinging, we know from pp. 90–91 that, when adjusting, the boat must be stopped *exactly* on the various headings north, south, east and west. To do this the corresponding relative bearings must be found from the list (Table A.1 or A.2) and the boat warped until the *exact* relative bearing is read either on the pelorus or on the sextant. It is for this reason that you must find a sheltered position and wait for a calm day. Otherwise keeping a boat on a datum heading can be most nerve-breaking and, furthermore, risk being inaccurate.

Once the boat is steady on the cardinal headings you proceed as indicated on p. 90.

WHEN TO REPEAT ADJUSTMENT

The compass should be readjusted at the beginning of every season, if important alterations are made to the boat, or if the boat undergoes a considerable change in latitude (of the order of magnitude of over 15°). It must also be remembered that, with time, the magnets used for compensation lose their magnetism.

CONCLUSION

Before finishing I should like to emphasise one point; many people believe that handling the compass is a kind of black art for a few initiated experts only, and, if I may be disrespectful, I would add that professional compass adjusters usually seem to do their best to perpetuate this belief. As a matter of fact adjusting a compass on a wooden yacht is really child's play: set the boat on heading north and put in transverse magnets until the compass reads 0°; set the boat on heading east and put in longitudinal magnets until the compass reads 90°; then check on headings 180° and 270° and halve the deviations which eventually appear. There is only one really difficult thing in the whole operation, but it has nothing to do with magnetism; it is warping the boat and keeping her steady, but if you are a good seaman you will certainly find it not too difficult and be able to do it accurately.

INDEX

Admiralty Chart Agents 8, 9, 77
Admiralty Chart Catalogue 8, 17
Admiralty charts 7, 10
Admiralty Hydrographic Department 8, 20, 75, 76, 79
Admiralty List of Lights 11, 18, 19, 75, 78
Admiralty List of Radio Signals 16, 17, 18, 19–20, 75, 78
Admiralty List of Radio Stations 17
Admiralty Notice to Mariners 10, 29
Admiralty Tidal Stream Atlases 19, 21, 22, 28, 34, 36
Almanacs. *See Brown's*; *Reed's*

Barograph 18
Barometer 18, 21, 57
Beacons. *See* Radio Beacons
Bearings, hand bearing-compass 6, 48
   magnetic 7–8, 93
   radio 5, 16–17, 46, 48–49, 51, 70, 71, 72, 79
   steady 35–36
   true 7–8, 13
   visual 46–48, 49, 51
Beating 36, 39–40, 57–63
Beme loop 17
Bini, Mario 3, 80
Books 18–20
Brown's Almanac 11, 18–19
Buoys 11
   light 10, 73

Celestial navigation 18, 48, 52, 69, 71, 72, 76
Chart instruments 12–16
Chart lockers 12
Charts 8–11, 21, 72, 74
   abbreviations 10–11
   consol 17, 78
   corrections 10
   foreign 7, 75
   gnomonic projection 77
   metric 8–9
   pilot 76
   plotting 79
   radio information 20
   study of 10–11
   tidal 34
   use in route planning 22–27
Chronometer 79
Compasses 1–8, 21
   adjusting 2–3, 95, 96, 97–98
   buying 2
   care of 5–6
   checking 2–4, 21, 37
   component deviations 83–84
      annulling 90
   correction card 8
   deviation 2, 8, 37, 80–81
      on wooden yachts 8, 90–91
   deviation card 2, 8, 21, 37
   deviation curve and its analysis 82–83
   effect of magnets and soft iron 84–89
   errors of 37
   for steel yachts 5
   hand bearing- 3–4, 6, 16, 21, 37
   magnetic 1–8
   magnetic variation 7–8
   mounting 2
   on Heron wireless 16
   standard 1
   steering 1–2, 3–4, 37
   swinging 2–4, 81, 93–95, 97
   tell-tale 7, 39
Component deviations. *See* Compasses
Consol radio navigation aid 16, 17, 46, 50, 72, 79
Currents 36, 70, 76

Dead reckoning 31–45, 69, 79
   allowance for tides 32–36
   checking and correction 46–55
   errors and their correction 32

Dead reckoning—*continued*
  not recorded 41
  reliability of plot 42–45
  sources of error 45
Deck watch 18, 79
Deviation. *See* Compasses
Dividers 16, 21, 22, 34

Echo sounders 17, 21, 46, 49, 70, 71, 72
Electrical equipment 16–17

Fixes, reliability of 50–52
Fog 44, 49, 69–72
Fog signals 11, 19
  synchronised 54
Foreign waters 75–79

Gnomonic projection chart 77
Great circle 76–78

Harbours 72–74
  foreign waters 75
Heeling error 2, 92–93
Heeling error instrument 93
Helmsman's error 38–40
Heron wireless 16
Hurst plotter 12–13

Instruments, chart 12–16
  plotting 12–13
Iron, hard 80–81, 84–89
  intermediate 80–81, 91–92
  soft 80–81, 83–89

Landfalls 4, 69
Lead 17, 18, 21, 49–50, 71
Leeway 36
Light-buoys 10, 73
Light-vessels 73
Lights 6, 19, 48, 72–74
  flashing characteristics 11, 28
  foreign waters 75
Line 17, 18, 21
Lines of position 45–55
  transferring 52–53
Log 17–18, 21, 31–32, 72
  sample 42
  writing up 41
Log book 21
Log error 37
Loxodrome 76–77

Magnetic bearings 7–8, 93
Magnetic declination 7
Magnetic heading 93–96
Magnetic rose 7, 37
Magnetic variation 7–8, 12, 14, 37
Magnetism, Magnets 2, 3, 5, 80–81, 83, 84–89, 91
Man overboard 54–55
Mariner's Handbook 20
Mercator's projection 76, 77
Meteorological services 17, 57
Meteorology 20, 56

Navigation, in other people's boats 20–21
  need for accuracy 30–31
  Also see Celestial Navigation
Navigational set-squares 7, 13–14, 21
Needles light 74

Ocean crossings 76–79
Ocean Passages for the World 76

Parallax 2, 37
Parallel rulers 7, 12, 21
Pelorus 94–95, 98
Pilot charts 76
Pilotage 72–74
Pilots 8, 20, 46, 69, 72, 75–76, 78
Planning and preparation 22–29
Plotting charts 79
Plotting instruments 12–13
Position, how to find 31
  lines of 45–46
    transferring 52–53
  need for accurate knowledge of 30–31, 54–55
Propeller error 41

Radio beacons 16, 20, 29, 46, 48–49, 51
Radio bearings. *See* Bearings
Radio information charts 20
Radio navigational aid, long-range *See* Consol
Reaching 40
Reed's Almanac 4, 11, 16, 17, 18–19, 21, 29
Rhumb-line 77
Road maps 11
Route, planning the 22–27
Rulers 21
Running 40, 63–64

Sea Traffic Separation Routes 29
Sestrel Navigator 21
Set-squares. *See* Navigational set-
    squares
Sextant 18, 79, 95, 98
Sextant angle, distance off by 54
Speedometer error 40
Stars. *See* Celestial navigation
Steady bearings. *See* Bearings
Steel yachts. *See* Yachts
Stop watch 18
Strategy 56–68
Sun. *See* Celestial navigation

Tacking 58–62
Tidal atlases. *See* Admiralty Tidal
    Stream Atlases
Tidal charts 34
Tides 22–24, 29, 64–68, 74
    allowance for 32–35
    times of 23, 28–29

True bearings. *See* Bearings

Uniform Buoyage System 11

Visibility, bad 17, 69–72
Visual bearings. *See* Bearings

Walker's log 17–18
Weather 56–57
Weather forecasts 16, 17
Wind 56–57
Wireless 16, 21, 79
Wooden yachts. *See* Yachts
Wreck-marking system 11

Yachts, steel 5, 16
    wooden 16, 83, 90–91